Reflection groups and Coxeter groups

T0275802

In this graduate textbook Professor Humphreys presents a concrete and up-to-date introduction to the theory of Coxeter groups. He assumes that the reader has a good knowledge of algebra, but otherwise the book is self-contained making it suitable either for courses and seminars or for self-study.

The first part is devoted to establishing concrete examples. Chapter 1 develops the most important facts about finite reflection groups and related geometry, leading to the presentation of such groups as Coxeter groups. In Chapter 2 these groups are classified by Coxeter graphs, and actual realizations are described. Chapter 3 discusses in detail the polynomial invariants of finite reflection groups. The first part ends with the construction in Chapter 4 of the affine Weyl groups, a class of Coxeter groups which plays a major role in Lie theory.

The second part (which is logically independent of, but motivated by, the first) starts by developing from scratch the properties of Coxeter groups in general, including the Bruhat ordering. In Chapter 6, it is shown how earlier examples and others fit into the general classification of Coxeter graphs. Chapter 7 introduces the seminal work of Kazhdan and Lusztig on representations of Hecke algebras associated with Coxeter groups. Finally, Chapter 8 sketches a number of interesting complementary topics as well as connections with Lie theory.

The book concludes with an extensive bibliography on Coxeter groups and their applications.

Already published

Reflection groups and Coxeter groups

JAMES E. HUMPHREYS

Professor of Mathematics

University of Massachusetts, Amherst

CAMBRIDGE
UNIVERSITY PRESS

CAMBRIDGE UNIVERSITY PRESS
Cambridge, New York, Melbourne, Madrid, Cape Town, Singapore,
São Paulo, Delhi, Dubai, Tokyo, Mexico City

Cambridge University Press
The Edinburgh Building, Cambridge CB2 8RU, UK

Published in the United States of America by
Cambridge University Press, New York

www.cambridge.org
Information on this title: www.cambridge.org/9780521436137

© Cambridge University Press 1990

First published 1990
First paperback edition (with corrections) 1992, 1994, 1997

A catalogue record for this publication is available from the British Library

Library of Congress Cataloguing in Publication Data

ISBN 978-0-521-37510-8 Hardback
ISBN 978-0-521-43613-7 Paperback

Contents

Preface

'Les choses, en effet, sont pour le moins doubles.'
Proust, *La Fugitive*

Since its appearance in 1968, Bourbaki [1] (treating Coxeter groups, Tits systems, reflection groups, and root systems) has become indispensable to all students of semisimple Lie theory. An enormous amount of information is packed into relatively few pages, including detailed descriptions of the individual root systems and a vast assortment of challenging 'exercises'. My own dog-eared copy (purchased at Dillon's in London in the spring of 1969 for 90 shillings) is always at hand. The present book attempts to be both an introduction to Bourbaki and an updating of the coverage, by inclusion of such topics as Bruhat ordering of Coxeter groups. I was motivated especially by the seminal 1979 paper of D.A. Kazhdan and G. Lusztig [1], which has led to rapid progress in representation theory and which deserves to be regarded as a fundamental chapter in the theory of Coxeter groups.

Part I deals concretely with two of the most important types of Coxeter groups: finite (real) reflection groups and affine Weyl groups. The treatment is fairly traditional, including the classification of associated Coxeter graphs and the detailed study of polynomial invariants of finite reflection groups.

Part II is for the most part logically independent of Part I, but lacks motivation without it. Chapter 5 develops the general theory of Coxeter groups, with emphasis on the 'root system' (following Deodhar [4]), the Strong Exchange Condition of Verma, and the Bruhat ordering. Special cases such as finite and hyperbolic Coxeter groups occupy Chapter 6. Chapter 7 is mainly an exposition of Kazhdan–Lusztig [1]. Finally, Chapter 8 sketches some related topics of interest, with suggestions for further reading. Because the subject reaches out in so many directions, I have provided an extensive (though by no means complete) bibliography.

The arguments in Part I are largely self-contained. However, the treatments of crystallographic reflection groups (Weyl groups) in Chapter 2 and affine Weyl groups in Chapter 4 require some facts about (crystallographic) root systems which are less directly connected with the theory of Coxeter groups and are therefore only summarized here. The coverage in Chapter VI of Bourbaki is thorough and accessible, and highly recommended for the serious student of Lie theory.

There are interesting groups generated by 'reflections' which are not in a natural way Coxeter groups, including for example most of the complex reflection groups (these deserve a book of their own). I have mentioned such related theories only in passing, in order to concentrate the treatment on Coxeter groups.

The history of the subject is long and intricate: see the *Note historique* in Bourbaki as well as the historical remarks in Coxeter [1]. Often a result has been first observed empirically (using the classification of finite reflection groups, for example) and later proved conceptually. I have tried to attribute theorems correctly, but have stopped short of reconstructing the history of each. The notes and references at the ends of chapters are intended to make it possible for the interested reader to get back to the original sources, notably the pioneering work of Coxeter and Witt. I hope readers will call omissions or errors to my attention.

All cross-references are to sections, such as 2.7. Each section contains at most one result labelled lemma, proposition, theorem or corollary, later referred to as (for example) Theorem 2.7. In order to emphasize what I take to be the high points in the development, I have made a distinction (admittedly subjective) between the labels 'proposition' and 'theorem'. Against considerable odds, I have struggled to make consistent notational choices, but there are occasional local aberrations. Exercises are scattered throughout the text. The reader is encouraged to try all of them; but none is required afterwards except as indicated.

I am indebted to the many people whose books, papers, and lectures have shaped my own knowledge of the subject, especially N. Bourbaki, V.V. Deodhar and J. Tits. Special thanks are due to George Avrunin for initiating me into the mysteries of LaTeX. Research support from the National Science Foundation is also gratefully acknowledged.

J.E. Humphreys
Amherst, MA
October 1989

For this printing, a number of misprints and minor errors have been corrected, and portions of 4.5, 5.5, 5.10 have been rewritten. I am grateful to the many readers who pointed out errors and suggested improvements, especially J. B. Carrell, E. Neher, and L. Tan.

Part I

Finite and affine reflection groups

Chapter 1

Finite reflection groups

In this chapter we begin the study of finite groups generated by reflections in (real) euclidean spaces. Our main tool will be a well-chosen set of vectors ('roots') orthogonal to reflecting hyperplanes (1.2). A set of 'simple roots' (1.3) yields an efficient generating set for the group (1.5), leading eventually to a very simple presentation by generators and relations as a 'Coxeter group' (1.9). The latter part of the chapter treats a number of geometric and group-theoretic topics, all of which involve the 'parabolic' subgroups generated by sets of simple reflections (1.10), e.g., Poincaré polynomials (1.11), fundamental domains (1.12), and the Coxeter complex (1.15).

1.1 Reflections

Recall what is meant by a **reflection** in a (real) euclidean space V endowed with a positive definite symmetric bilinear form (λ, μ). A reflection is a linear operator s on V which sends some nonzero vector α to its negative while fixing pointwise the hyperplane H_α orthogonal to α. We may write $s = s_\alpha$, bearing in mind however that $s_\alpha = s_{c\alpha}$ for any nonzero $c \in \mathbf{R}$. There is a simple formula:

$$s_\alpha \lambda = \lambda - \frac{2(\lambda, \alpha)}{(\alpha, \alpha)} \alpha.$$

Indeed, this is correct when $\lambda = \alpha$ and when $\lambda \in H_\alpha$; so it is correct for all $\lambda \in V = \mathbf{R}\alpha \oplus H_\alpha$. A quick calculation (left to the reader) shows that s_α is an orthogonal transformation, i.e., $(s_\alpha \lambda, s_\alpha \mu) = (\lambda, \mu)$ for all $\lambda, \mu \in V$. It is clear that $s_\alpha^2 = 1$, so s_α has order 2 in the group $O(V)$ of all orthogonal transformations of V.

A finite group generated by reflections (or **finite reflection group**, for short) is an especially interesting type of finite subgroup of $O(V)$.

The purpose of this chapter and the next will be to classify and describe all such groups. In doing so, we shall explore alternately the internal structure of the group itself (e.g., the relations satisfied by the generating reflections) and the geometric aspects of the action of the group on V (e.g., fundamental domains).

Here are some basic examples, which should be kept in mind as the story unfolds. (They are labelled by 'types', in accordance with the classification to be carried out in Chapter 2.)

$(I_2(m), m \geq 3)$ Take V to be the euclidean plane, and define $\mathcal{D}_m$ to be the **dihedral group** of order $2m$, consisting of the orthogonal transformations which preserve a regular m-sided polygon centered at the origin. $\mathcal{D}_m$ contains m rotations (through multiples of $2\pi/m$) and m reflections (about the 'diagonals' of the polygon). Here 'diagonal' means a line bisecting the polygon, joining two vertices or the midpoints of opposite sides if m is even, or joining a vertex to the midpoint of the opposite side if m is odd. Note that the rotations form a cyclic subgroup of index 2, generated by a rotation through $2\pi/m$. The group $\mathcal{D}_m$ is actually generated by reflections, because a rotation through $2\pi/m$ can be achieved as a product of two reflections relative to a pair of adjacent diagonals which meet at an angle of $\theta := \pi/m$ (see Figure 1). Let

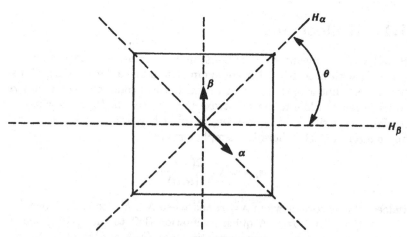

Figure 1: The case $m = 4$

the reflecting lines H_α and H_β contain these diagonals, and choose the orthogonal unit vectors $\alpha = (\sin\theta, -\cos\theta)$ and $\beta = (0, 1)$ which form an obtuse angle of $\pi - \theta$, so $(\alpha, \beta) = -\cos\theta$. To see that $s_\alpha s_\beta$ is a (counterclockwise) rotation through 2θ, take H_β to be the x-axis and compute with 2×2 matrices relative to the standard basis of $\mathbf{R}^2$:

$$\begin{pmatrix} \cos 2\theta & \sin 2\theta \\ \sin 2\theta & -\cos 2\theta \end{pmatrix} \begin{pmatrix} 1 & 0 \\ 0 & -1 \end{pmatrix} = \begin{pmatrix} \cos 2\theta & -\sin 2\theta \\ \sin 2\theta & \cos 2\theta \end{pmatrix}$$

Exercise 1. The reflections form a single conjugacy class in $\mathcal{D}_m$ when m is odd, but form two classes when m is even.

$(A_{n-1}, \ n \geq 2)$ Consider the **symmetric group** $\mathcal{S}_n$. It can be thought of as a subgroup of the group $O(n, \mathbf{R})$ of $n \times n$ orthogonal matrices in the following way. Make a permutation act on $\mathbf{R}^n$ by permuting the standard basis vectors $\varepsilon_1, \ldots, \varepsilon_n$ (permute the subscripts). Observe that the transposition (ij) acts as a reflection, sending $\varepsilon_i - \varepsilon_j$ to its negative and fixing pointwise the orthogonal complement, which consists of all vectors in $\mathbf{R}^n$ having equal ith and jth components. Since $\mathcal{S}_n$ is generated by transpositions, it is a reflection group. Indeed, it is already generated by the transpositions $(i, i+1), 1 \leq i \leq n-1$.

Exercise 2. Regarding $\mathcal{S}_n$ in this way as a subgroup of $O(n, \mathbf{R})$, prove that the transpositions are the sole reflections belonging to $\mathcal{S}_n$.

When $\mathcal{S}_n$ acts on $\mathbf{R}^n$ in the way just described, it fixes pointwise the line spanned by $\varepsilon_1 + \ldots + \varepsilon_n$ (these are clearly the only fixed points) and leaves stable the orthogonal complement, the hyperplane consisting of vectors whose coordinates add up to 0. Thus $\mathcal{S}_n$ also acts on an $(n-1)$-dimensional euclidean space as a group generated by reflections, fixing no point except the origin. This accounts for the subscript $n-1$ in the label A_{n-1}. When a reflection group W acts on V with no nonzero fixed points, we say that W is **essential** relative to V. It is clear that any subgroup W of $O(V)$ stabilizes the orthogonal complement V' of its space of fixed points and is essential relative to V'.

$(B_n, \ n \geq 2)$ Again let $V = \mathbf{R}^n$, so $\mathcal{S}_n$ acts on V as above. Other reflections can be defined by sending an ε_i to its negative and fixing all other ε_j. These sign changes generate a group of order 2^n isomorphic to $(\mathbf{Z}/2\mathbf{Z})^n$, which intersects $\mathcal{S}_n$ trivially and is normalized by $\mathcal{S}_n$: conjugating the sign change $\varepsilon_i \mapsto -\varepsilon_i$ by a transposition yields another such sign change. Thus the semidirect product of $\mathcal{S}_n$ and the group of sign changes yields a reflection group W of order $2^n n!$. It is easy to check that W is essential.

$(D_n, \ n \geq 4)$ We can get another reflection group acting on $\mathbf{R}^n$, a subgroup of index 2 in the group of type B_n just described: $\mathcal{S}_n$ clearly normalizes the subgroup consisting of sign changes which involve an *even* number of signs, generated by the reflections $\varepsilon_i + \varepsilon_j \mapsto -(\varepsilon_i + \varepsilon_j), i \neq j$. So the semidirect product is also a reflection group (and is essential).

1.2   Roots

From now on we denote by W a finite reflection group, acting on the euclidean space V. The letter W is used because 'most' finite reflection groups turn out to be 'Weyl groups' (associated with semisimple Lie algebras or Lie groups). Much of the theory to be developed in this book is in fact motivated by the problems of Lie theory, cf. Bourbaki [1].

In order to understand the internal structure of W as an abstract group, we first explore the way in which W acts on V. Each reflection s_α in W determines a reflecting hyperplane H_α and a line $L_\alpha = \mathbf{R}\alpha$ orthogonal to it. The following result implies that W permutes the collection of all such lines.

Proposition   *If $t \in O(V)$ and α is any nonzero vector in V, then $ts_\alpha t^{-1} = s_{t\alpha}$. In particular, if $w \in W$, then $s_{w\alpha}$ belongs to W whenever s_α does.*

Proof. Obviously $ts_\alpha t^{-1}$ sends $t\alpha$ to its negative. So we need only show that $ts_\alpha t^{-1}$ fixes $H_{t\alpha}$ pointwise. Note that λ lies in H_α if and only if $t\lambda$ lies in $H_{t\alpha}$, since $(\lambda, \alpha) = (t\lambda, t\alpha)$. In turn, $(ts_\alpha t^{-1})(t\lambda) = ts_\alpha\lambda = t\lambda$ whenever λ lies in H_α. $\square$

Thus W permutes the lines L_α, where s_α ranges over the set of reflections contained in W, via $w(L_\alpha) = L_{w\alpha}$. Only the lines L_α are determined by W, not the vectors α. However, if we select the pairs of unit vectors lying in all such lines, the collection of vectors so obtained will be stable under the action of W. It is this sort of geometric configuration which we shall emphasize below. Actually, we need not insist that the vectors be of equal length: only the stability under W is significant for our purposes. For example, the dihedral group $\mathcal{D}_4$ preserves the collection of eight vectors in $\mathbf{R}^2$:

$$\pm(1,0), \ \pm(1,1), \ \pm(0,1), \ \pm(-1,1)$$

For flexibility in some future arguments, it is most convenient to axiomatize the situation as follows. Take Φ to be a finite set of nonzero vectors in V satisfying the conditions:

(R1) $\Phi \cap \mathbf{R}\alpha = \{\alpha, -\alpha\}$ for all $\alpha \in \Phi$;
(R2) $s_\alpha \Phi = \Phi$ for all $\alpha \in \Phi$.

Then define W to be the group generated by all reflections $s_\alpha, \alpha \in \Phi$. Call Φ a **root system** with associated reflection group W. The elements of Φ are called **roots** because of the historical connection between Weyl groups and semisimple Lie algebras, where the notion of 'root' goes back ultimately to the characteristic roots of certain operators on the Lie algebra. However, our notion of 'root system' differs somewhat from that encountered in Lie theory; see 2.9 below.

As the previous discussion shows, any finite reflection group can be realized in this way, possibly for many different choices of Φ. Conversely, any group W arising from a root system is in fact finite. Indeed, each $s_\alpha(\alpha \in \Phi)$ and hence each element of W fixes pointwise the orthogonal complement of the subspace spanned by Φ. So only $w = 1$ can fix all elements of Φ. This means that the natural homomorphism of W into the symmetric group on Φ has trivial kernel, forcing W to be finite.

To recapitulate: our finite reflection group $W \subset O(V)$ is henceforth to be studied in conjunction with a root system $\Phi \subset V$, subject only to (R1) and (R2) above. The choice of Φ is somewhat flexible. It might consist of unit vectors, or not. The reflections $s_\alpha(\alpha \in \Phi)$ might or might not be known to exhaust all reflections in W. The set Φ might span V, or not. All that really matters for later arguments is that (R1) and (R2) hold.

Remark. Given a root system Φ and corresponding reflection group W, define Φ' to be the set of unit vectors proportional to the vectors in Φ. Then Φ' is clearly a root system, with W as corresponding reflection group.

1.3 Positive and simple systems

Fix a root system Φ in the euclidean space V, so that W is the finite reflection group generated by all $s_\alpha(\alpha \in \Phi)$. While W is completely determined by the geometric configuration Φ, there is one serious drawback to using Φ as a tool in the classification of possible reflection groups: Φ may be extremely large compared with the dimension of V. For example, when W is a dihedral group, Φ may have just as many elements as W, even though $\dim V = 2$.

This leads us to look for a linearly independent subset of Φ (a 'simple system') from which Φ can somehow be reconstituted. More precisely, we ask that each root be an $\mathbf{R}$-linear combination of 'simple' roots with coefficients all of like sign. In this way a simple system will yield a partition of Φ into 'positive' and 'negative' roots, with precisely one of each pair $\{\alpha, -\alpha\}$ labelled as positive. Partitions of this sort are easy to find (by totally ordering V), so we take this as our starting point in the search for a simple system.

Recall that a **total ordering** of the real vector space V is a transitive relation on V (denoted $<$) satisfying the following axioms.

(1) For each pair $\lambda, \mu \in V$, exactly one of $\lambda < \mu, \lambda = \mu, \mu < \lambda$ holds.

(2) For all λ, μ, ν in V, if $\mu < \nu$, then $\lambda + \mu < \lambda + \nu$.

(3) If $\mu < \nu$ and c is a nonzero real number, then $c\mu < c\nu$ if $c > 0$, while $c\nu < c\mu$ if $c < 0$.

Given such an ordering, we say that $\lambda \in V$ is **positive** if $0 < \lambda$. The sum of positive vectors is positive, as is the scalar multiple of a positive vector by a positive real number.

To construct a total ordering of V is easy: choose an arbitrary ordered basis $\lambda_1, \ldots, \lambda_n$ of V and adopt the corresponding lexicographic order, where $\sum a_i \lambda_i < \sum b_i \lambda_i$ means that $a_k < b_k$ if k is the least index i for which $a_i \neq b_i$. The reader can quickly verify the axioms above. Note too that all λ_i are positive in this ordering.

Returning to the root system Φ, we call a subset Π a **positive system** if it consists of all those roots which are positive relative to some total ordering of V. It is clear that positive systems exist. Moreover, since roots come in pairs $\{\alpha, -\alpha\}$, it is clear that Φ must be the disjoint union of Π and $-\Pi$, the latter being called a **negative system**. When Π is fixed, we can write $\alpha > 0$ in place of $\alpha \in \Pi$.

Call a subset Δ of Φ a **simple system** (and call its elements **simple roots**) if Δ is a vector space basis for the **R**-span of Φ in V and if moreover each $\alpha \in \Phi$ is a linear combination of Δ with coefficients all of the same sign (all nonnegative or all nonpositive). It is not at all evident that simple systems exist.

Theorem   (a) *If Δ is a simple system in Φ, then there is a unique positive system containing Δ.*

(b) *Every positive system Π in Φ contains a unique simple system; in particular, simple systems exist.*

Proof. (a) Suppose the simple system Δ is contained in a positive system Π. Then all roots which are nonnegative linear combinations of Δ must also be in Π (and their negatives cannot be in Π). So Π is characterized uniquely as the set of all such roots. To see that such a positive system exists, extend the linearly independent set Δ to an ordered basis of V and take Π to be the set of positive elements of Φ in the corresponding lexicographic ordering. Evidently $\Delta \subset \Pi$.

(b) Suppose for a moment that the given positive system Π (coming from some total ordering of V) does contain a simple system Δ. Then Δ may be characterized as the set of all roots in $\alpha \in \Pi$ such that α is not expressible as a linear combination with strictly positive coefficients of two or more elements of Π. (This follows easily from the definitions.) So Δ is the unique simple system in Π.

How can we actually locate a simple system in Π? Choose as small a subset $\Delta \subset \Pi$ as possible subject to the requirement that each root in Π be a nonnegative linear combination of Δ. Obviously such a subset exists. We need only prove that Δ is linearly independent. This will follow from a key geometric condition, to be verified below:

$$(\alpha, \beta) \leq 0 \text{ for all pairs } \alpha \neq \beta \text{ in } \Delta. \qquad (1)$$

Assuming the truth of (1), consider what would happen if Δ failed to be linearly independent: $\sum_{\alpha \in \Delta} a_\alpha \alpha = 0$, with not all $a_\alpha = 0$. Rewrite this as $\sum b_\beta \beta = \sum c_\gamma \gamma$, where the sums are taken over disjoint subsets of Δ and the coefficients are strictly positive. If σ denotes the sum just written, we have $\sigma > 0$. But, thanks to (1),

$$0 \leq (\sigma, \sigma) = \left(\sum b_\beta \beta, \sum c_\gamma \gamma \right) \leq 0.$$

This forces $\sigma = 0$, which is absurd. Thus Δ must be linearly independent.

It remains to verify (1). Suppose it fails for some pair α, β. Then the formula for a reflection gives $s_\alpha \beta = \beta - c\alpha$, with $c = 2(\beta, \alpha)/(\alpha, \alpha) > 0$. Since $s_\alpha \beta \in \Phi$, either it or its negative must lie in Π. Say $s_\alpha \beta = \sum c_\gamma \gamma$ (sum over $\gamma \in \Delta, c_\gamma \geq 0$). In case $c_\beta < 1$, we get $s_\alpha \beta = \beta - c\alpha = c_\beta \beta + \sum_{\gamma \neq \beta} c_\gamma \gamma$, or $(1 - c_\beta)\beta =$ nonnegative linear combination of $\Delta \setminus \{\beta\}$. Since $1 - c_\beta > 0$, this allows us to discard β, contradicting the minimality of Δ. In case $c_\beta \geq 1$, we get instead $0 = (c_\beta - 1)\beta + c\alpha + \sum_{\gamma \neq \beta} c_\gamma \gamma$. But a nonnegative linear combination of Δ with at least one positive coefficient cannot equal 0, by definition of total ordering. So $s_\alpha \beta$ cannot be positive. A similar argument shows that $s_\alpha \beta$ cannot be negative either; here the cases to consider are $c + c_\alpha > 0$ and $c + c_\alpha \leq 0$. This contradiction implies that (1) must be true. $\square$

Because of the uniqueness statements in the theorem, the proof actually shows that (1) must hold for any simple system. This is an important geometric constraint, which plays a role in the classification of possible reflection groups (Chapter 2):

Corollary (of proof) *If Δ is a simple system in Φ, then $(\alpha, \beta) \leq 0$ for all $\alpha \neq \beta$ in Δ.* $\square$

The cardinality of any simple system is an invariant of Φ, since it measures the dimension of the span of Φ in V. We call it the **rank** of W. For example, $\mathcal{D}_m$ has rank 2, while $\mathcal{S}_n$ has rank $n - 1$.

Exercise 1. If Φ has rank 2, prove that W is a dihedral group. [This will be easier to do after Theorem 1.5.]

Exercise 2. Find simple systems for the various groups described in 1.1, taking for Φ in each case a convenient set of vectors (not necessarily unit vectors).

1.4   Conjugacy of positive and simple systems

We have shown that positive and simple systems in Φ determine each other uniquely. However, we have not ruled out the unpleasant possibility that differently chosen simple systems might differ drastically as geometric configurations. Here we examine the relationship between different systems.

It follows directly from the definition that, for any simple system Δ and for any $w \in W$, $w\Delta$ is again a simple system, with corresponding positive system $w\Pi$ (if Π is the positive system determined by Δ). To understand better the passage from Π to $w\Pi$, consider the special case $w = s_\alpha$ ($\alpha \in \Delta$). We find that Π and $s_\alpha\Pi$ differ only by one root:

Proposition   *Let Δ be a simple system, contained in the positive system Π. If $\alpha \in \Delta$, then $s_\alpha(\Pi\backslash\{\alpha\}) = \Pi\backslash\{\alpha\}$.*

Proof. Let $\beta \in \Pi, \beta \neq \alpha$, and write $\beta = \sum_{\gamma \in \Delta} c_\gamma \gamma$ (with all $c_\gamma \geq 0$). Since the only multiples of α in Φ are $\pm\alpha$, some $c_\gamma > 0$ for $\gamma \neq \alpha$. Now apply s_α to both sides: $s_\alpha\beta = \beta - c\alpha$ is a linear combination of Δ involving γ with the same coefficient c_γ. Because all coefficients in such an expression have like sign, $s_\alpha\beta$ must be positive. It cannot be α, for then we reach the contradiction: $\beta = s_\alpha s_\alpha\beta = s_\alpha\alpha = -\alpha$ (which is not in Π). Thus s_α maps $\Pi\backslash\{\alpha\}$ into itself (injectively), hence onto itself. $\square$

Besides being the key step in the proof of the theorem below, this result is often helpful in recognizing when a root is in fact equal to a given simple root α: it characterizes α as the sole positive root made negative by s_α.

Theorem   *Any two positive (resp. simple) systems in Φ are conjugate under W.*

Proof. Let Π and Π' be positive systems, so each contains precisely half of the roots. Proceed by induction on $r = \text{Card}(\Pi \cap -\Pi')$. If $r = 0$, then $\Pi = \Pi'$ and we are done. If $r > 0$, then clearly the simple system Δ in Π cannot be wholly contained in Π'. Choose $\alpha \in \Delta$ with $\alpha \in -\Pi'$. The proposition above implies that $\text{Card}(s_\alpha\Pi \cap -\Pi') = r-1$. Induction, applied to the positive systems $s_\alpha\Pi$ and Π', furnishes an element $w \in W$ for which $w(s_\alpha\Pi) = \Pi'$. $\square$

1.5   Generation by simple reflections

Fix a simple system Δ and corresponding positive system Π in Φ. (Theorem 1.4 shows that it makes no great difference which Δ we choose.) Our next goal is to show that W is generated by **simple reflections**, i.e., those s_α for which $\alpha \in \Delta$. First, a definition: if $\beta \in \Phi$, write uniquely $\beta = \sum_{\alpha \in \Delta} c_\alpha \alpha$, and call $\sum c_\alpha$ the **height** of β (relative to Δ), abbreviated ht(β). For example, ht(β) = 1 if $\beta \in \Delta$.

Theorem *For a fixed simple system Δ, W is generated by the reflections $s_\alpha (\alpha \in \Delta)$.*

Proof. Denote by W' the subgroup of W so generated. We proceed in several steps to show that $W' = W$.

(1) If $\beta \in \Pi$, consider $W'\beta \cap \Pi$. This is a nonempty set of positive roots (containing at least β), and we can choose from it an element γ of smallest possible height. We claim that $\gamma \in \Delta$. Write $\gamma = \sum_{\alpha \in \Delta} c_\alpha \alpha$, and note that $0 < (\gamma, \gamma) = \sum c_\alpha (\gamma, \alpha)$, forcing $(\gamma, \alpha) > 0$ for some $\alpha \in \Delta$. If $\gamma = \alpha$, we are satisfied. Otherwise consider the root $s_\alpha \gamma$, which is positive according to Proposition 1.4. Since $s_\alpha \gamma$ is obtained from γ by subtracting a positive multiple of α, we have ht($s_\alpha \gamma$) < ht(γ). But $s_\alpha \gamma \in W'\beta$ (since $s_\alpha \in W'$), contradicting the original choice of γ. So indeed $\gamma = \alpha$ must be simple.

(2) Now we can argue that $W'\Delta = \Phi$. We just showed that the W'-orbit of any positive root β meets Δ, so that $\Pi \subset W'\Delta$. On the other hand, if β is negative, then $-\beta \in \Pi$ is conjugate by some $w \in W'$ to some $\alpha \in \Delta$. Then $-\beta = w\alpha$ forces $\beta = (ws_\alpha)\alpha$, with $ws_\alpha \in W'$. Thus $-\Pi \subset W'\Delta$.

(3) Finally, take any generator s_β of W. Use step (2) to write $\beta = w\alpha$ for some $w \in W'$ and some $\alpha \in \Delta$. Then Proposition 1.2 shows that $s_\beta = ws_\alpha w^{-1} \in W'$. This proves that $W = W'$. $\square$

A useful byproduct of the proof is the fact that every root can attain the status of a simple root (relative to some positive system):

Corollary **(of proof)** *Given Δ, for every $\beta \in \Phi$ there exists $w \in W$ such that $w\beta \in \Delta$.* $\square$

Exercise 1. Let Φ be a root system of rank n consisting of unit vectors. If $\Psi \subset \Phi$ is a set of n roots whose mutual angles agree with those between the roots in some simple system, then Ψ must be a simple system.

Exercise 2. Given a simple system Δ, no proper subset of the simple reflections can generate W. [Otherwise find $\alpha \in \Delta$ for which s_α is not needed as a generator of W. Consider $w \in W$ for which $w(-\alpha) \in \Delta$.]

Exercise 3. If $\beta \in \Pi \backslash \Delta$, prove that ht($\beta$) > 1.

Having seen that W can be generated by relatively few reflections, we may go on to seek an efficient presentation of W as an abstract group, using these generators together with suitable relations. Certain relations among the $s_\alpha (\alpha \in \Delta)$ are obvious: those of the form

$$(s_\alpha s_\beta)^{m(\alpha,\beta)} = 1,$$

where $m(\alpha, \beta)$ denotes the order of the product in W. It turns out (Theorem 1.9 below) that these obvious relations completely determine W. This is not difficult to verify in the case of $\mathcal{D}_m$, but is already rather challenging in the case of $\mathcal{S}_n$ (try it!).

1.6   The length function

In order to obtain the promised presentation of W, we need to study closely the way in which an arbitrary $w \in W$ can be written as a product of simple reflections, say $w = s_1 \cdots s_r$ (where $s_i = s_{\alpha_i}$ for some $\alpha_i \in \Delta$). Define the **length** $\ell(w)$ of w (relative to Δ) to be the smallest r for which such an expression exists, and call the expression **reduced**. By convention, $\ell(1) = 0$.

Clearly $\ell(w) = 1$ if and only if $w = s_\alpha$ for some $\alpha \in \Delta$. It is also clear that $\ell(w) = \ell(w^{-1})$, since $w^{-1} = s_r \cdots s_1$ implies $\ell(w^{-1}) \leq \ell(w)$, and vice versa. Another easy property of the length function follows from the fact that each reflection has determinant -1 as a linear operator:

$$\det(w) = (-1)^{\ell(w)}.$$

Indeed, $\det(w) = (-1)^r$ whenever w can be written as a product of r reflections, so any such r has the same parity as $\ell(w)$. From this it follows that $\ell(ww')$ has the same parity as $\ell(w) + \ell(w')$. In particular, if $\ell(w) = r$ and $\alpha \in \Delta$, then $\ell(s_\alpha w)$ is either $r + 1$ or $r - 1$.

It will be shown in 1.7 that $\ell(w)$ can be characterized geometrically as the number of positive roots sent by w to negative roots. In case $w = s_\alpha (\alpha \in \Delta)$, this is the content of Proposition 1.4. Here we lay some of the groundwork.

Having fixed Δ and the corresponding positive system Π, define $n(w) := \mathrm{Card}(\Pi \cap w^{-1}(-\Pi)) = $ number of positive roots sent to negative roots by w. Observe that $n(w^{-1}) = n(w)$, because $\Pi \cap w^{-1}(-\Pi) = w^{-1}(w\Pi \cap -\Pi) = -w^{-1}(\Pi \cap w(-\Pi))$, which has the same number of elements as $\Pi \cap w(-\Pi)$. This is reassuring, given the similar property of the length function.

Lemma   *Let $\alpha \in \Delta, w \in W$. Then:*
   (a) $w\alpha > 0 \Rightarrow n(ws_\alpha) = n(w) + 1$.
   (b) $w\alpha < 0 \Rightarrow n(ws_\alpha) = n(w) - 1$.
   (c) $w^{-1}\alpha > 0 \Rightarrow n(s_\alpha w) = n(w) + 1$.
   (d) $w^{-1}\alpha < 0 \Rightarrow n(s_\alpha w) = n(w) - 1$.

Proof. Set $\Pi(w) := \Pi \cap w^{-1}(-\Pi)$, so that $n(w) = \mathrm{Card}\,\Pi(w)$. If $w\alpha > 0$, observe that $\Pi(ws_\alpha)$ is the disjoint union of $s_\alpha\Pi(w)$ and $\{\alpha\}$, thanks to Proposition 1.4. If $w\alpha < 0$, the same result implies that $s_\alpha \Pi(ws_\alpha) = \Pi(w)\backslash\{\alpha\}$, whereas α does lie in $\Pi(w)$. This establishes (a) and (b). To get (c) and (d), replace w by w^{-1} and use the fact that $n(w^{-1}s_\alpha) = n(s_\alpha w)$. $\square$

Corollary *If $w \in W$ is written in any way as a product of simple reflections, say $w = s_1 \cdots s_r$, then $n(w) \leq r$. In particular, $n(w) \leq \ell(w)$.*

Proof. As we build up the expression for w in r steps, the value of the n function (initially 0) can increase by at most 1 at each step, according to the lemma. $\square$

Some further properties of the length function which are shared by the n function are described in the following exercise.

Exercise 1. (a) If $w \in W$, prove that $\det(w) = (-1)^{n(w)}$. (b) If $w, w' \in W$, prove that $n(ww') \leq n(w) + n(w')$ and $n(ww') \equiv n(w) + n(w')(\bmod 2)$.

Exercise 2. Taking the simple reflections in $\mathcal{S}_n$ to be the transpositions $(i, i+1)$, show that the length of a permutation π is the number of 'inversions': the number of pairs $i < j$ for which $\pi(i) > \pi(j)$.

1.7 Deletion and Exchange Conditions

The following crucial result reveals how a product of simple reflections may be shortened if it is not already as short as possible.

Theorem *Fix a simple system Δ. Let $w = s_1 \cdots s_r$ be any expression of $w \in W$ as a product of simple reflections (say $s_i = s_{\alpha_i}$, with repetitions permitted). Suppose $n(w) < r$. Then there exist indices $1 \leq i < j \leq r$ satisfying:*
   (a) $\alpha_i = (s_{i+1} \cdots s_{j-1})\alpha_j,$
   (b) $s_{i+1}s_{i+2} \cdots s_j = s_i s_{i+1} \cdots s_{j-1},$
   (c) $w = s_1 \cdots \hat{s}_i \cdots \hat{s}_j \cdots s_r$ *(where the hat denotes omission).*

Proof. (a) Because $n(w) < r$, iteration of part (a) of Lemma 1.6 shows that, for some $j \leq r$, we have $(s_1 \cdots s_{j-1})\alpha_j < 0$. But since $\alpha_j > 0$, there is an index $i < j$ for which $s_i(s_{i+1} \cdots s_{j-1})\alpha_j < 0$ while $(s_{i+1} \cdots s_{j-1})\alpha_j > 0$. (In case $i = j - 1, s_{i+1} \cdots s_{j-1}$ is interpreted to be 1.) Now Proposition 1.4, applied to the simple reflection s_i, implies that the positive root $(s_{i+1} \cdots s_{j-1})\alpha_j$ made negative by s_i must be α_i.

(b) Set $\alpha = \alpha_j, w' = s_{i+1} \cdots s_{j-1}$, so that $w'\alpha = \alpha_i$ by part (a). By Proposition 1.2, $w's_\alpha w'^{-1} = s_{w'\alpha} = s_i$, which means that

$$(s_{i+1} \cdots s_{j-1})s_j(s_{j-1} \cdots s_{i+1}) = s_i.$$

Multiply both sides on the right by $s_{i+1} \cdots s_{j-1}$ to get the desired identity.

(c) This is just another way to express (b): multiply both sides of (b) on the right by s_j to obtain $s_{i+1} \cdots s_{j-1} = s_i \cdots s_j$, and then substitute in the original expression for w. □

Corollary *If $w \in W$, then $n(w) = \ell(w)$.*

Proof. By Corollary 1.6, $n(w) \le \ell(w)$. Suppose $n(w) < \ell(w) = r$, and write $w = s_1 \cdots s_r$ (reduced expression). Then part (c) above allows us to rewrite w as the product of $r - 2$ simple reflections, contrary to $\ell(w) = r$. □

Having identified the functions ℓ and n, we can restate Lemma 1.6: multiplying w on the right by $s_\alpha(\alpha \in \Delta)$ increases the length by 1 if $w\alpha > 0$ and decreases the length by 1 if $w\alpha < 0$, etc.

We can also reinterpret part (c) of the theorem, which may be called the **Deletion Condition**: given an expression $w = s_1 \cdots s_r$ which is not reduced, there exist indices $1 \le i < j \le r$ such that $w = s_1 \cdots \hat{s}_i \cdots \hat{s}_j \cdots s_r$. Thus successive omissions of pairs of factors will eventually yield a reduced expression.

To get a better feel for what the corollary says, it is useful to ask how we might enumerate for a given w the set $\Pi(w)$. Since Card $\Pi(w) = n(w) = \ell(w)$, the answer ought to have something to do with the nature of a reduced expression $w = s_1 \cdots s_r(s_i = s_{\alpha_i})$. Indeed, given such an expression, consider the r roots

$$\beta_i := s_r s_{r-1} \cdots s_{i+1}(\alpha_i), \text{ with } \beta_r := \alpha_r.$$

We claim that $\Pi(w) = \{\beta_1, \ldots, \beta_r\}$, where the β_i are distinct. To see this, let $\beta \in \Pi(w)$. Since $\beta > 0$ but $w\beta < 0$, we can find an index $i \le r$ such that $(s_{i+1} \cdots s_r)\beta > 0$ while $(s_i s_{i+1} \cdots s_r)\beta < 0$; in case $i = r$, interpret $s_{i+1} \cdots s_r$ as 1. Thus the positive root $(s_{i+1} \cdots s_r)\beta$ is sent by s_i to a negative root; Proposition 1.4 forces $(s_{i+1} \cdots s_r)\beta = \alpha_i$, whence $\beta = \beta_i$. As a result, $\Pi(w) \subset \{\beta_1, \ldots, \beta_r\}$. Because Card $\Pi(w) = r$, equality must hold (and the β_i must be distinct).

There is a nice way to reformulate the essence of the theorem:

Exchange Condition *Let $w = s_1 \cdots s_r$ (not necessarily reduced), where each s_i is a simple reflection. If $\ell(ws) < \ell(w)$ for some simple reflection $s = s_\alpha$, then there exists an index i for which $ws = s_1 \cdots \hat{s}_i \cdots s_r$ (and thus $w = s_1 \cdots \hat{s}_i \cdots s_r s$, with a factor s exchanged for a factor s_i). In particular, w has a reduced expression ending in s if and only if $\ell(ws) < \ell(w)$.*

Proof. The hypothesis $\ell(ws) < \ell(w)$ is now known to be equivalent to: $w\alpha < 0$. Repeating the proof of the above theorem for the expression

$ws = s_1 \cdots s_r s$, we can therefore take $j = r + 1$ in part (a) and then conclude in part (c) that

$$ws = s_1 \cdots \hat{s}_i \cdots s_r.$$

This yields the desired expression for w. □

Remark. In Bourbaki [1] (and elsewhere in the literature), the Exchange Condition is stated under the stricter condition that the given expression for w is *reduced*. This is easily seen to be equivalent to our Deletion Condition. It is clearly implied by the Deletion Condition. In the other direction, let $w = s_1 \cdots s_r$ (not reduced). Say $s_1 \cdots s_{j-1}$ is reduced, but $s_1 \cdots s_j$ is not. By the stricter version of the Exchange Condition, there is an index $i < j$ for which $s_1 \cdots s_{j-1} = s_1 \cdots \hat{s}_i \cdots s_j$, whence $w = s_1 \cdots \hat{s}_i \cdots \hat{s}_j \cdots s_r$, as required. It is less obvious that our Exchange Condition is a formal consequence of Bourbaki's (our derivation being based on the way W acts on roots). This turns out to be true, as the result of some indirect arguments given in the setting of general Coxeter groups; see 5.8 below. The reader might want to attempt a more direct line of argument.

Exercise 1. In the Exchange Condition, suppose $\ell(w) = r$. Prove that the index i in the conclusion is uniquely determined.

Exercise 2. Formulate a 'left-handed' version of the Exchange Condition, under the hypothesis $\ell(sw) < \ell(w)$.

1.8 Simple transitivity and the longest element

Theorem 1.4 expressed the fact that W permutes the various positive (or simple) systems in a transitive fashion. Corollary 1.7 immediately implies the following result, which shows that the permutation action of W is simply transitive.

Theorem *Let Δ be a simple system, Π the corresponding positive system. The following conditions on $w \in W$ are equivalent:*
  (a) $w\Pi = \Pi$;
  (b) $w\Delta = \Delta$;
  (c) $n(w) = 0$;
  (d) $\ell(w) = 0$;
  (e) $w = 1$. □

One corollary of the simple transitivity is well worth exploring. It is clear from the definition that $-\Pi$ is a positive system whenever Π is. So there must exist a unique element $w_o \in W$ sending Π to $-\Pi$. Moreover,

$\ell(w_o) = n(w_o) = \text{Card II}$ is as large as possible, and no other element of W has as great a length. In particular, $w_o^{-1} = w_o$. Using Lemma 1.6 we can characterize w_o as the unique $w \in W$ satisfying $\ell(ws_\alpha) < \ell(w)$ for all $\alpha \in \Delta$. This has an interesting consequence. Given a reduced expression $w = s_1 \cdots s_r$, we can successively multiply w on the right by simple reflections (increasing the length by 1) until this is no longer possible and w_o is obtained. Thus $w_o = ww'$ with $\ell(w_o) = \ell(w) + \ell(w')$ for some $w' \in W$. The conclusion can also be reformulated as follows:

$$\ell(w_o w) = \ell(w_o) - \ell(w) \text{ for all } w \in W. \tag{2}$$

Exercise 1. What is w_o in the case of S_n, relative to the simple system $\varepsilon_1 - \varepsilon_2, \ldots, \varepsilon_{n-1} - \varepsilon_n$?

Exercise 2. In any reduced expression for w_o, every simple reflection must occur at least once.

1.9 Generators and relations

Now we are prepared to verify the presentation of W described at the end of 1.5. Recall that $m(\alpha, \beta)$ denotes the order of $s_\alpha s_\beta$ in W, for any roots α, β. For example, $m(\alpha, \alpha) = 1$. (We could also write $m(s_\alpha, s_\beta)$.)

Theorem *Fix a simple system Δ in Φ. Then W is generated by the set $S := \{s_\alpha,\ \alpha \in \Delta\}$, subject only to the relations:*

$$(s_\alpha s_\beta)^{m(\alpha, \beta)} = 1 \ (\alpha, \beta \in \Delta).$$

Proof. Rather than introduce notation for a free group on a set having the cardinality of Δ, with normal subgroup generated by appropriate words in the free generators, we argue informally that each relation in W is a consequence of the given relations. It has to be shown that each relation

$$s_1 \cdots s_r = 1 \text{ (where } s_i = s_{\alpha_i} \text{ for some } \alpha_i \in \Delta) \tag{3}$$

is a consequence of the given relations. Note that r must be even, since $\det(s_i) = -1$. If $r = 2$, the equation reads $s_1 s_2 = 1$, forcing $s_1 = s_2^{-1} = s_2$ because of the relation $s_2^2 = 1$. So (3) becomes $s_1^2 = 1$, one of the given relations. Proceed now by induction on $r = 2q$, and let $q > 1$. The relations $s_i^2 = 1$ will henceforth be used tacitly whenever needed to rewrite expressions. For example, (3) can be rewritten as

$$s_{i+1} \cdots s_r s_1 \cdots s_i = 1. \tag{4}$$

We will repeatedly invoke the Deletion Condition (1.7). Apply it first to the element

$$s_1 \cdots s_{q+1} = s_r \cdots s_{q+2}.$$

Since the length of the right side is at most $q - 1$, the left side cannot be a reduced expression. Part (b) of Theorem 1.7 (which is equivalent to the Deletion Condition) then yields indices $1 \leq i < j \leq q + 1$ for which

$$s_{i+1} \cdots s_j = s_i \cdots s_{j-1}, \tag{5}$$

which is equivalent to the relation

$$s_i \cdots s_{j-1} s_j \cdots s_{i+1} = 1. \tag{6}$$

In case (6) involves fewer than r simple reflections, the induction hypothesis says that it can be derived from the given relations. Then it is permissible to replace $s_{i+1} \cdots s_j$ by $s_i \cdots s_{j-1}$ in (3) and rewrite (3) as

$$s_1 \cdots s_i (s_i \cdots s_{j-1}) s_{j+1} \cdots s_r = s_1 \cdots \hat{s}_i \cdots \hat{s}_j \ldots s_r = 1.$$

Again by induction, this last relation is a consequence of the given ones; so the same is true of (3). The only catch comes when (6) still involves r simple reflections: then $i = 1, j = q + 1$, and (5) becomes

$$s_2 \cdots s_{q+1} = s_1 \cdots s_q. \tag{7}$$

We could attempt to avoid this impasse by using another version (4) of our original relation (3), say

$$s_2 \cdots s_r s_1 = 1.$$

Repetition of the above steps will now be successful unless

$$s_3 \cdots s_{q+2} = s_2 \cdots s_{q+1}. \tag{8}$$

In the presence of both (7) and (8), a different strategy is needed. If we can just show that (8) is a consequence of the given relations, we can substitute it in (3) and conclude as before. Rewritten once more, (8) becomes

$$s_3 (s_2 s_3 \cdots s_{q+1}) s_{q+2} s_{q+1} \cdots s_4 = 1.$$

The left side is a product of r simple reflections, just like (3), so we can again try our original line of argument. This will be successful unless

$$s_2 \cdots s_{q+1} = s_3 s_2 s_3 \cdots s_q. \tag{9}$$

But (9) and (7) together force $s_1 = s_3$. Similarly, we could cyclically permute factors and reach a successful conclusion unless $s_2 = s_4$. Continuing step-by-step in this way (another induction!), we reach a total impasse only in case $s_1 = s_3 = \ldots = s_{r-1}$ and $s_2 = s_4 = \ldots = s_r$. But then (3) has the form

$$s_\alpha s_\beta s_\alpha s_\beta \cdots s_\alpha s_\beta = 1,$$

which is a consequence of the given relation $(s_\alpha s_\beta)^{m(\alpha,\beta)} = 1.\square$

The presentation of W just obtained is about as simple as could be hoped for. Any group (finite or infinite) having such a presentation relative to a generating set S is called a **Coxeter group**; more precisely, the pair (W, S) is called a **Coxeter system**. It is required that all $m(\alpha, \alpha) = 1$, but a relation $(s_\alpha s_\beta)^{m(\alpha,\beta)} = 1$ may be omitted to allow the product to have infinite order. Part II will be devoted to the detailed study of this rather large class of groups. Eventually (in Chapter 6) it will be seen that the finite Coxeter groups are precisely the finite reflection groups.

While the proof of the theorem is still fresh in the reader's mind, we point out that the steps depend formally just on the Deletion Condition. (This fact will be invoked in Chapter 4.) It will be seen in Chapter 5 that groups which satisfy a condition of this type are essentially the same thing as Coxeter groups, so our choice of strategy in the present proof was not accidental.

1.10   Parabolic subgroups and minimal coset representatives

Let us pause to take stock of where we are in our study of finite reflection groups. We have been studying such a group W in tandem with a root system, which leads to a small set of generating reflections (corresponding to a simple system). The simple system must be an independent set of vectors at mutually obtuse angles, constrained strongly by the fact that distinct pairs of reflections generate finite dihedral groups. In Chapter 2 we shall classify the possible geometric configurations of this sort and thereby classify the groups.

Meanwhile, we want to explore further the subgroup structure of W, in conjunction with various geometric features of the action of W on V. Much of what we do in the remainder of this chapter will be essential in the latter half of Chapter 3; but the results are also of interest in themselves.

We begin by looking more closely at the subgroups of W generated by sets of simple reflections (for any fixed simple system Δ). In order to be consistent with the notation to be introduced in Chapter 5, we label these subgroups as follows. Having fixed Δ, let S be the set of simple reflections $s_\alpha, \alpha \in \Delta$. For any subset $I \subset S$, define W_I to be the subgroup of W generated by all $s_\alpha \in I$, and let $\Delta_I := \{\alpha \in \Delta | s_\alpha \in I\}$. At the extremes, $W_\emptyset = \{1\}$ and $W_S = W$. Replacing Δ by another simple system $w\Delta$ would just replace W_I by its conjugate $wW_I w^{-1}$. All subgroups of W obtainable in this way are called **parabolic subgroups** (for somewhat arcane reasons which we won't attempt to explain here). They arise constantly in the further study and application of reflection groups, in part because they facilitate inductive arguments.

Proposition *Fix a simple system Δ and the corresponding set S of simple reflections. Let $I \subset S$, and define Φ_I to be the intersection of Φ with the $\mathbf{R}$-span V_I of Δ_I in V.*

(a) Φ_I is a root system in V (resp. V_I), with simple system Δ_I and with corresponding reflection group W_I (resp. W_I restricted to V_I).

(b) Viewing W_I as a reflection group, with length function ℓ_I relative to the simple system Δ_I, we have $\ell = \ell_I$ on W_I.

(c) Define $W^I := \{w \in W | \ell(ws) > \ell(w)$ for all $s \in I\}$. Given $w \in W$, there is a unique $u \in W^I$ and a unique $v \in W_I$ such that $w = uv$. Their lengths satisfy $\ell(w) = \ell(u) + \ell(v)$. Moreover, u is the unique element of smallest length in the coset wW_I.

Proof. (a) It is clear that W_I stabilizes V_I and that conditions (R1) and (R2) in 1.2 are satisfied by Φ_I (viewed as a subset of either V or V_I). It is also clear that Δ_I is a simple system. Therefore the group W_I (acting on either V or V_I) is the corresponding reflection group.

(b) We invoke the characterization of the length function given in 1.7: $\ell(w)$ is the number of positive roots sent to negative roots by w, and similarly for ℓ_I (where the 'positive' roots relative to Δ_I are clearly those in $\Phi^+ \cap \Phi_I$). Now suppose $\alpha \in \Phi^+ \setminus \Phi_I$. Then α involves some simple root $\gamma \notin \Delta_I$, so for all $\beta \in \Delta_I$, $s_\beta \alpha$ still involves γ with a positive coefficient. It follows that $s_\beta \alpha > 0$. In turn, for all $w \in W_I$, we get $w\alpha > 0$. Thus the roots in Φ^+ sent by $w \in W_I$ to negative roots are precisely the roots in Φ_I^+ sent by w to negative roots. This means $\ell(w) = \ell_I(w)$.

(c) Given $w \in W$, choose a coset representative $u \in wW_I$ of smallest possible length, and write $w = uv$ for $v \in W_I$. Since $us \in wW_I$ for all $s \in I$, it is clear that $u \in W^I$. Now write reduced expressions: $u = s_1 \cdots s_q$ ($s_i \in S$) and $v = s_1' \cdots s_r'$ (where we may assume $s_i' \in I$, thanks to (b)). Then $\ell(w) \leq \ell(u) + \ell(v) = q + r$. If the inequality were strict, the Deletion Condition (1.7) would allow us to omit two of the factors s_i or s_i' in uv without changing w. But omitting any factor from u would yield a coset representative in wW_I of smaller length than u, contrary to the choice we made. So two factors s_i', s_j' can be omitted without changing v, contrary to the fact that the expression for v is reduced. Therefore $\ell(w) = \ell(u) + \ell(v)$.

The only fact about w used in this argument is that it belongs to the coset wW_I; so we have actually shown that any element of this coset can be written in the form uv, with $\ell(uv) = \ell(u) + \ell(v)$. Here u is a fixed coset representative of smallest length (forcing $u \in W^I$). In particular, u is the unique coset representative of smallest length.

Suppose there were another element $u' \in W^I$ lying in wW_I, with $u' \neq u$. We could then write $u' = uv$ with $\ell(v) = r > 0$, say $v = s_1 \cdots s_r$ ($s_i \in I$). But then $\ell(u's_r) < \ell(u')$ contrary to $u' \in W^I$. $\square$

The distinguished coset representatives W^I in part (c) of the proposition may be called **minimal coset representatives**. They will play an essential role in the following section, as well as in 1.15.

Exercise 1. Is there a result analogous to (c) describing minimal representatives of the double cosets $W_I w W_J$ ($I, J \subset S$)?

Exercise 2. When $W = S_n$, prove that each parabolic subgroup of W is isomorphic to a direct product of symmetric groups.

Exercise 3. Given $s \neq s'$ in S, set $v := ss'ss' \cdots$ (m factors, where m is the order of ss' in W), so also $v = s'ss's \cdots$ (m factors), and $v^2 = 1$. If $w \in W$ satisfies $\ell(ws) < \ell(w)$ and $\ell(ws') < \ell(w)$, prove that $\ell(wv) = \ell(w) - m$. [Consider W_I, $I = \{s, s'\}$. Since $w = (wv)v$ with $v \in W_I$, it suffices to show that $wv \in W^I$. Look at the action on the roots corresponding to s, s'.]

1.11 Poincaré polynomials

Part (c) of Proposition 1.10 has a nice application to the study of the 'growth' of W relative to the generating set S. This is measured by the sequence
$$a_n := \text{Card}\,\{w \in W \,|\, \ell(w) = n\},$$
which in turn defines a polynomial in the indeterminate t:
$$W(t) := \sum_{n \geq 0} a_n t^n = \sum_{w \in W} t^{\ell(w)}.$$

For example, when $W = S_3$, $W(t) = 1 + 2t + 2t^2 + t^3$. Because of its homological interpretation in special cases (see the remark at the end of 3.15), we refer to $W(t)$ as the **Poincaré polynomial** of W.

More generally, for an arbitrary subset $X \subset W$, we can define
$$X(t) := \sum_{w \in X} t^{\ell(w)}.$$

Note for example that for $I \subset S$, $W_I(t)$ coincides with the Poincaré polynomial of the reflection group W_I (since ℓ agrees with the length function ℓ_I). It is an immediate consequence of part (c) of Proposition 1.10 that

$$W(t) = W_I(t)W^I(t).$$

This can be used to derive an effective algorithm for computing $W(t)$ by induction on $|S|$. For brevity, write $(-1)^I$ instead of $(-1)^{|I|}$. Recall (1.8) that W has a unique element w_0 of maximum length $N := |\Pi|$.

Proposition

$$\sum_{I \subseteq S}(-1)^I \frac{W(t)}{W_I(t)} = \sum_{I \subseteq S}(-1)^I W^I(t) = t^N.$$

Proof. The equality of the first and second sum follows from the above remarks. In turn, consider the contribution which a fixed $w \in W$ makes to the second sum. Set $K := \{s \in S | \ell(ws) > \ell(w)\}$. Then $w \in W^I$ precisely when $I \subset K$, so $t^{\ell(w)}$ occurs in the sum with co-efficient $\sum_{I \subseteq K}(-1)^I$. Unless K is empty, it is an easy combinatorial exercise to show that this quantity is 0. But $K = \emptyset$ precisely when $w = w_0$, thereby accounting for the surviving term t^N on the right. $\square$

Exercise 1. When $W = S_3$, use the formula in the proposition to compute $W(t)$ inductively, starting with the fact that $W(t) = 1 + t$ for a group of rank 1. Do the same for $W = \mathcal{D}_m$ in general.

Note that when 1 is substituted for t, $W_I(t)$ becomes $|W_I|$. So the formula in the proposition yields an identity (due originally to Witt [1], Satz 3):

$$\sum_{I \subseteq S}(-1)^I \frac{|W|}{|W_I|} = 1.$$

Exercise 2. The identity just obtained permits an inductive calculation of $|W|$ when $|S|$ is *odd*. Suppose for example that $|S| = 3$, and that the dihedral subgroups W_I are of respective orders 4, 6, 10. What is $|W|$?

1.12 Fundamental domains

The goal of this section (and the ones following) is to refine the description of the action of W on V (in terms of orbits and isotropy groups), with emphasis on the role of the reflecting hyperplanes. In the process we get a nice geometric interpretation of the simple transitivity of W on simple systems, as well as further information about parabolic subgroups.

Fix a positive system Π, containing the simple system Δ. Associated with each hyperplane H_α are the open half-spaces A_α and A'_α, where

$$A_\alpha := \{\lambda \in V | (\lambda, \alpha) > 0\}$$

and $A'_\alpha := -A_\alpha$. Define $C := \bigcap_{\alpha \in \Delta} A_\alpha$. As an intersection of open convex sets, C is itself open and convex. It is also a cone (closed under positive scalar multiples). Let D be the closure $\bar{C}$, the intersection of closed half-spaces $H_\alpha \cup A_\alpha$. Thus

$$D = \{\lambda \in V | (\lambda, \alpha) \geq 0 \text{ for all } \alpha \in \Delta\}.$$

Clearly D is a closed convex cone. We intend to show that D is a **fundamental domain** for the action of W on V, i.e., each $\lambda \in V$ is conjugate under W to one and only one point in D. One part of this is straightforward:

Lemma   *Each $\lambda \in V$ is W-conjugate to some $\mu \in D$. Moreover, $\mu - \lambda$ is a nonnegative $\mathbf{R}$-linear combination of Δ.*

Proof. Introduce a partial ordering of V (not to be confused with earlier total orderings, which are no longer needed): $\lambda \leq \mu$ if and only if $\mu - \lambda$ is a linear combination of Δ with nonnegative coefficients. It is trivial to verify that this is a partial ordering. Consider those W-conjugates μ of λ which satisfy $\lambda \leq \mu$. From this nonempty set (which contains at least λ), choose a maximal element μ. If $\alpha \in \Delta, s_\alpha\mu$ is obtained from μ by subtracting a multiple of α, namely $2(\mu, \alpha)/(\alpha, \alpha)$. Since this is another W-conjugate of λ, the maximality of μ forces $(\mu, \alpha) \geq 0$. This holds for all $\alpha \in \Delta$, so $\mu \in D$ as desired. $\square$

To see that each λ is W-conjugate to at most one $\mu \in D$, it is enough to show that no pair of distinct elements of D can be W-conjugate. In the course of the proof, we can get some sharper information about the **isotropy group** $\{w \in W | w\mu = \mu\}$ for an arbitrary $\mu \in V$.

Theorem   *Fix $\Pi \supset \Delta$ (hence D), as above.*

(a) *If $w\lambda = \mu$ for $\lambda, \mu \in D$, then $\lambda = \mu$ and w is a product of simple reflections fixing λ. In particular, if $\lambda \in C$, then the isotropy group of λ is trivial.*

(b) *D is a fundamental domain for the action of W on V.*

(c) *If $\lambda \in V$, the isotropy group of λ is generated by those reflections s_α ($\alpha \in \Phi$) which it contains.*

(d) *If U is any subset of V, then the subgroup of W fixing U pointwise is generated by those reflections s_α which it contains.*

Proof. (a) Proceed by induction on $\ell(w) = n(w)$. If $n(w) = 0$, then $w = 1$ and there is nothing to prove. If $n(w) > 0$, then w must send some simple root α to a negative root (otherwise $w\Delta$ and hence $w\Pi$ would consist of positive roots). Thanks to part (b) of Lemma 1.6, $n(ws_\alpha) = n(w) - 1$. Moreover, since $\lambda, \mu \in D$, with $w\alpha < 0$, we have: $0 \geq (\mu, w\alpha) = (w^{-1}\mu, w^{-1}w\alpha) = (\lambda, \alpha) \geq 0$, which forces $(\lambda, \alpha) = 0$ and $s_\alpha\lambda = \lambda$. Therefore $ws_\alpha\lambda = \mu$. By induction, $\lambda = \mu$ and ws_α is a product of simple reflections fixing λ; so w is also such a product.

(b) This follows at once from part (a), together with the above lemma.

(c) Given $\lambda \in V$, use the lemma to find $w \in W$ for which $\mu := w\lambda$ lies in D. By part (a), the isotropy group W' of μ is generated by the simple reflections it contains. It is clear that $w^{-1}W'w$ is the isotropy group W' of λ. Since conjugates of simple reflections are again reflections with respect to roots, it follows that W' is generated by those s_α which it contains.

(d) The subgroup W^0 fixing pointwise the span of U (or a basis $\lambda_1, \ldots, \lambda_t$ of this span) is clearly the same as the subgroup fixing U pointwise. In turn, W^0 is just the intersection of the isotropy groups of the λ_i. Proceed by induction on t, the case $t = 1$ being settled by part (c). We know that the isotropy group W' of λ_1 is generated by the set of all reflections s_α which it contains (α running over a subset Φ' of Φ containing pairs of roots $\alpha, -\alpha$). Proposition 1.2 implies that W' stabilizes Φ'. So this reflection group (with root system Φ') can take the place of W. By induction (formulated to cover all possible reflection groups!), its subgroup fixing $\{\lambda_2, \ldots, \lambda_t\}$ pointwise is generated by some of the reflections s_α, $\alpha \in \Phi'$. But this subgroup is just W^0. $\square$

Exercise 1. Show how the theorem can be used to solve the *Word Problem* for W: given a product of simple reflections, decide whether or not it equals 1 in W.

Exercise 2. If $U \subset D$ in part (d) of the theorem, then the subgroup of W fixing U pointwise is generated by simple reflections.

Exercise 3. If $w \in W$ is an involution (an element of order 2), prove that w can be written as a product of commuting reflections. [Use induction on the dimension of V.]

We have now associated with each simple system Δ an open convex cone C in V whose points all have trivial isotropy group in W. It is clear that replacing Δ by $w\Delta$ just replaces C by wC. Thus the simply transitive action (1.8) of W on simple systems translates into a simply transitive action on this family of open sets, which we call **chambers**. The chambers are characterized topologically as the connected components of the complement in V of $\bigcup_\alpha H_\alpha$. Given a chamber C corresponding to a simple system Δ, its **walls** are defined to be the hyperplanes H_α ($\alpha \in \Delta$). Each wall has a 'positive' and a 'negative' side (with C lying on the positive side). Then the roots in Δ can be characterized as those roots which are orthogonal to some wall of C and positively directed. Note finally that the angle between any two walls of a chamber is an angle of the form π/k for a positive integer $k \geq 2$. This follows from our discussion of dihedral groups in 1.1.

Exercise 4. Prove that $\ell(w)$ equals the number of hyperplanes $H_\alpha (\alpha > 0)$ which separate C from wC.

1.13   The lattice of parabolic subgroups

We can use part (d) of Theorem 1.12 to obtain a clearer picture of the collection of parabolic subgroups of the form W_I, where I runs over the subsets of the set S of simple reflections relative to a fixed choice of Δ.

Proposition   *Under the correspondence $I \mapsto W_I$, the collection of parabolic subgroups W_I $(I \subset S)$ is isomorphic to the lattice of subsets of S.*

Proof. It is clear that $W_{I \cup J}$ is the group generated by W_I and W_J $(I, J \subset S)$. We claim that $W_I \cap W_J = W_{I \cap J}$, from which it will follow immediately that the map $I \mapsto W_I$ is one-to-one and defines a lattice isomorphism. Only one inclusion is obvious: $W_{I \cap J} \subset W_I \cap W_J$.

Recall from 1.10 the subspaces V_I and V_J of V. It is clear from the definition that $V_I \cap V_J = V_{I \cap J}$. Now recall from linear algebra the fact that, for any two subspaces $A, B \subset V$, $(A \cap B)^\perp = A^\perp + B^\perp$ (proved by comparison of dimensions). From this we get:

$$V_I^\perp + V_J^\perp = (V_I \cap V_J)^\perp = V_{I \cap J}^\perp.$$

Now suppose $w \in W_I \cap W_J$, so w fixes each vector in $V_I^\perp + V_J^\perp = V_{I \cap J}^\perp$. According to part (d) of Theorem 1.12, w is a product of reflections s_α which also fix this space pointwise. But then each such α is orthogonal to this space, hence lies in $\Phi \cap V_{I \cap J} = \Phi_{I \cap J}$. It follows that $w \in W_{I \cap J}$ as required. $\square$

Exercise. If $s_1, \ldots, s_r$ are distinct elements of S, then $\ell(s_1 \cdots s_r) = r$.

1.14   Reflections in W

Theorem 1.12 also helps to clear up a possible ambiguity in the way W is specified. Recall that our study of W has depended on a fixed choice of a root system Φ, with W defined as the group generated by the reflections s_α $(\alpha \in \Phi)$. There was no requirement that these s_α should exhaust the reflections in W. But this turns out to be true anyway.

Proposition   *Every reflection in W is of the form s_α for some $\alpha \in \Phi$.*

Proof. Let s be a reflection in W, with reflecting hyperplane H fixed pointwise by s. Thus s lies in the isotropy group of H, which is nontrivial and thus (thanks to part (d) of Theorem 1.12) is generated by some of the reflections s_α $(\alpha \in \Phi)$. But s_α cannot fix H pointwise unless $H = H_\alpha$, in which case $s = s_\alpha$. $\square$

Exercise. Let Ψ be any subset of Φ for which the reflections s_α ($\alpha \in \Psi$) generate W. Prove that every $\alpha \in \Phi$ is W-conjugate to some element of Ψ. [Consider $\Phi' := \{w\alpha | w \in W, \alpha \in \Psi\}$. This set satisfies the axioms for a root system (1.2), with W as the associated reflection group.]

1.15 The Coxeter complex

We can give a more detailed description of the fundamental domain D in 1.12 in terms of parabolic subgroups. As before, fix a simple system Δ and corresponding set S of simple reflections. It is convenient to assume that Δ spans V. For each subset I of S, define

$$C_I := \{\lambda \in D | (\lambda, \alpha) = 0 \text{ for all } \alpha \in \Delta_I, (\lambda, \alpha) > 0 \text{ for all } \alpha \in \Delta \setminus \Delta_I\}.$$

Thus C_I is an intersection of certain hyperplanes H_α and certain open half-spaces A_α. It is clear that the sets C_I partition D, with $C_\emptyset = C$ and $C_S = \{0\}$. Moreover, the linear span of C_I has dimension $n - |I|$, where $n = \dim V$.

Thanks to Theorem 1.12, V is partitioned by the collection $\mathcal{C}$ of all sets $wC_I (w \in W, I \subset S)$. More precisely, for each fixed I the sets wC_I and $w'C_I$ are disjoint unless w and w' lie in the same left coset in W/W_I, in which case they coincide. For distinct I and J, all sets wC_I and $w'C_J$ are disjoint. We call $\mathcal{C}$ the **Coxeter complex** of W. Any set wC_I is called a **facet** of type I.

Proposition *For each $I \subset S$, the isotropy group of the facet C_I of $\mathcal{C}$ is precisely W_I. Thus the parabolic subgroups of W are the isotropy groups of the elements of $\mathcal{C}$.*

Proof. From the definition of C_I, it is clear that W_I fixes it pointwise. Now suppose $w \in W$ satisfies: $wC_I = C_I$. By part (a) of Theorem 1.12, w fixes C_I pointwise.

Use part (c) of Proposition 1.10 to write $w = uv$, where $v \in W_I$ and u satisfies: $\ell(us_\alpha) > \ell(u)$ for all $\alpha \in I$. Thanks to 1.6, this condition implies that $u\Delta_I \subset \Phi^+$. If $u \neq 1$, there must be some $\alpha \in \Delta$ for which $u\alpha < 0$, and (as just observed) $\alpha \notin \Delta_I$. Choose any $\lambda \in C_I$, so $w\lambda = u\lambda = \lambda$. Since $\alpha \notin \Delta_I$, we have by definition: $(\lambda, \alpha) > 0$. On the other hand, $u\alpha < 0$ forces $(\lambda, \alpha) = (u\lambda, u\alpha) = (\lambda, u\alpha) \leq 0$, which is absurd. $\square$

The characterization of parabolic subgroups as isotropy groups yields an interpretation of $\mathcal{C}$ as an abstract simplicial complex: The 'vertices' are the left cosets wW_I, where I is maximal in S (obtained by discarding one simple reflection). A finite set of vertices determines a 'simplex' if these vertices (left cosets) have a nonempty intersection. The dimension of the complex is $n - 1$ (one less than the cardinality of the largest possible simplex).

When Δ spans V, $\mathcal{C}$ has a natural geometric realization: by intersecting its elements with the unit sphere in V, one gets a simplicial decomposition of the sphere. Because W preserves the simplicial structure, it also acts on the integral homology groups of the sphere. This leads to an interesting formula for the character det of W (realized on the top homology) in terms of the permutation characters of W on the cosets of parabolic subgroups, via the Hopf trace formula. In the following section we shall derive an algebraic version of this formula.

1.16   An alternating sum formula

In this section we obtain an alternating sum formula for $\det(w)$, which involves counting how many elements of each dimension are fixed by w in the Coxeter complex $\mathcal{C}$. This formula will be a key ingredient in 3.15, and could be deferred until then. (We present it here while the features of the Coxeter complex are still fresh in the reader's mind.) First we derive a general combinatorial formula, which the reader may recognize as an Euler characteristic computation.

Let $H_1, \ldots, H_r$ be an arbitrary collection of hyperplanes in the euclidean space V (of dimension n), and form a complex $\mathcal{K}$ in the same way we formed the Coxeter complex. Each hyperplane $H = H^0$ determines a positive half-space H^+ and a negative half-space H^-. Then a typical element of $\mathcal{K}$ is a (nonempty) intersection of the form

$$K = \bigcap H_i^{\varepsilon_i}, \quad \text{where } \varepsilon_i \in \{0, +, -\}.$$

We write $\dim K = i$ if the linear span has dimension i. Note that this linear span L is obtained by intersecting all H_i^0 which occur in the definition of K. In turn, K is the *open* subset of L obtained by intersecting various open half-spaces with L.

Lemma   *Denote by n_i the number of elements of $\mathcal{K}$ having dimension i. Then $\sum_i (-1)^i n_i = (-1)^n$.*

Proof. We use induction on the number r of hyperplanes used to define $\mathcal{K}$ (the case $r = 1$ being clear). What is the effect of adding to the list $H_1, \ldots, H_r$ a new hyperplane H? New elements of the complex are created just in case H intersects some K in a proper nonempty subset. Let L be the linear span of K in V. If $x \in H \cap K$, we can find an open neighborhood U of x in L contained in K, by the above remarks. Since $H \cap L$ has codimension 1 in L, it is clear that U meets both H^+ and H^-. Thus we replace the single element K by two new elements $H^+ \cap K, H^- \cap K$ of dimension i, together with an element $H^0 \cap K$ of dimension $i - 1$. This increases the original n_i and n_{i-1} by 1, leaving the alternating sum unchanged. $\square$

Fix Δ and S as before. Consider, for each $I \subset S$, the facets vC_I of type I; these are in bijective correspondence with left cosets vW_I. For $w \in W$, define $f_I(w)$ to be the number of such facets stabilized (i.e., fixed pointwise) by w. This is the same as the number of left cosets vW_I fixed under left multiplication by w. As before, we write $(-1)^I$ instead of $(-1)^{|I|}$.

Proposition

$$\sum_{I \subset S} (-1)^I f_I(w) = \det(w).$$

Proof. Fix w and let V' be the subspace of V fixed pointwise by w (the 1-eigenspace). Then the facets in C fixed by w are precisely those which lie in V'. Let K be the complex obtained by intersecting the elements of C with V'. Then the number n_i of facets of dimension i in C which lie in V' is just the number of facets of dimension i in K, so we can apply the above lemma to this situation, with $c := \dim V'$:

$$\sum_i (-1)^i n_i = (-1)^c.$$

If $n := \dim V$, $\dim C_I = n - |I|$, so we see that

$$n_i = \sum_{|I|=n-i} f_I(w).$$

Combining, we get

$$(-1)^n \sum_{I \subset S} (-1)^I f_I(w) = (-1)^c.$$

But w is an orthogonal transformation, so its possible eigenvalues are 1 (with multiplicity c), b pairs of complex conjugate numbers of absolute value 1, and -1 (with multiplicity $n - c - 2b$). Accordingly, $\det(w) = (-1)^{n-c-2b} = (-1)^{n-c}$. So the proposition follows. $\square$

Notes

We follow the approach in the Appendix to Steinberg [4], supplemented by Chapter 2 of Carter [1] (where the arguments are given only for Weyl groups, but usually remain valid for all finite reflection groups). See also Curtis–Reiner [3], §64. The treatment in Grove–Benson [1] (or the earlier edition, Benson–Grove [1]) is more leisurely, giving for example a detailed account of finite groups of orthogonal transformations in dimensions 2 and 3. For more of the geometry of finite reflection groups, see Coxeter [1], Chapter XI.

(1.9) This sort of presentation was apparently first studied systematically by Coxeter [2][3] and Witt [1]. Emphasis on the Exchange Condition came later, in Matsumoto [1], Iwahori–Matsumoto [1], Bourbaki [1], IV, §1.

(1.11) The proposition, due to Solomon [3], will be used in 3.15 below.

(1.15) Coxeter complexes (for general Coxeter groups) are studied in detail in Brown [1], Ronan [1]; this is directed to the study of 'buildings', as developed by Tits [1][6] (etc.). See also Carter [1], Chapters 2, 15.

(1.16) Solomon [3] gave a topological proof of the proposition; the version here is due to Steinberg [5], §2. It will be used in 3.15 below.

Chapter 2

Classification of finite reflection groups

The goal of this chapter is to determine all possible finite reflection groups, in terms of their 'Coxeter graphs' (2.3). Although a general existence proof will be given in Chapters 5 and 6, we shall describe in some detail how to construct each (irreducible) type of group and compute its order. Groups satisfying a crystallographic condition (2.8) are especially important in Lie theory, where they arise as Weyl groups and can be studied uniformly.

Throughout the chapter we reserve the index n for the *rank* of W.

2.1 Isomorphisms

The presentation of W obtained in Theorem 1.9 shows that (as an abstract group) W is determined up to isomorphism by the set of integers $m(\alpha, \beta)$, $\alpha, \beta \in \Delta$. A convenient way to encode this information in a picture is to construct a graph Γ with vertex set in one-to-one correspondence with Δ; join a pair of vertices corresponding to $\alpha \neq \beta$ by an edge whenever $m(\alpha, \beta) \geq 3$, and label such an edge with $m(\alpha, \beta)$. (For a pair of vertices not joined by an edge, it is then understood that $m(\alpha, \beta) = 2$.) This labelled graph is called the **Coxeter graph** of W. It determines W up to isomorphism. Since simple systems are conjugate, it does not depend on the choice of Δ.

For example, the graph of $\mathcal{D}_m$ is

$$\circ \overset{m}{\text{---}} \circ,$$

while the graph of $\mathcal{S}_{n+1}$ has n vertices:

$$\circ \overset{3}{\text{---}} \circ \cdots \circ \overset{3}{\text{---}} \circ.$$

The classification of finite reflection groups given in this chapter will rely heavily on the study of possible Coxeter graphs.

We can give a somewhat more precise criterion for reflection groups to be isomorphic, in the geometric setting:

Proposition *For $i = 1, 2$, let W_i be a finite reflection group acting on the euclidean space V_i. Assume W_i is essential. If W_1 and W_2 have the same Coxeter graph, then there is an isometry of V_1 onto V_2 inducing an isomorphism of W_1 onto W_2. (In particular, if $V_1 = V_2$, the subgroups W_1 and W_2 are conjugate in $O(V)$.)*

Proof. Fix a simple system Δ_i for W_i. By assumption, Δ_i is a basis of V_i. As remarked in 1.2, we may assume that all roots are of unit length. Let φ map Δ_1 to Δ_2 in a way compatible with the common Coxeter graph, and extend by linearity to a vector space isomorphism of V_1 onto V_2. If $\alpha \neq \beta$ lie in Δ_1, the angle θ between them is $\pi - \pi/m(\alpha, \beta)$. Since roots are unit vectors, we get $(\alpha, \beta) = \cos\theta = -\cos(\pi/m(\alpha, \beta))$. The same calculation applies to the inner product of the roots in Δ_2 corresponding to α, β, since the same $m(\alpha, \beta)$ occurs. Thus φ is an isometry, which clearly induces an isomorphism of W_1 onto W_2. $\square$

2.2   Irreducible components

We say that the Coxeter system (W, S) is **irreducible** if the Coxeter graph Γ is connected. (We also call Φ irreducible in this case.) In general, let $\Gamma_1, \ldots, \Gamma_r$ be the connected components of Γ, and let Δ_i, S_i be the corresponding sets of simple roots and simple reflections. Thus if $\alpha \in \Delta_i$ and $\beta \in \Delta_j$ ($i \neq j$), we have $m(\alpha, \beta) = 2$ and therefore $s_\alpha s_\beta = s_\beta s_\alpha$. The following proposition shows that the study of finite reflection groups can be largely reduced to the case when Γ is connected.

Proposition *Let (W, S) have Coxeter graph Γ, with connected components $\Gamma_1, \ldots, \Gamma_r$, and let $S_1, \ldots, S_r$ be the corresponding subsets of S. Then W is the direct product of the parabolic subgroups $W_{S_1}, \ldots, W_{S_r}$, and each Coxeter system (W_{S_i}, S_i) is irreducible.*

Proof. Use induction on r. Since the elements of S_i commute with the elements of S_j when $i \neq j$, it is clear that the indicated parabolic subgroups centralize each other, hence that each is normal in W. Moreover, the product of these subgroups contains S and therefore must be all of W. By induction, $W_{S \setminus S_i}$ is the direct product of the remaining W_{S_j}, and Proposition 1.13 implies that W_{S_i} intersects it trivially. So the product is direct. $\square$

Exercise. Let W be the dihedral group $\mathcal{D}_6$ of order 12, with standard Coxeter generators $S = \{s, s'\}$. The Coxeter system (W, S) is irreducible. However, W has another set S' of Coxeter generators leading to a Coxeter system which is not irreducible: $S' := \{s, (s's)^3, s(s's)^2\}$.

2.3 Coxeter graphs and associated bilinear forms

We start with a general definition, applicable not only to finite reflection groups but also to other Coxeter groups encountered later on. Define a **Coxeter graph** to be a finite (undirected) graph, whose edges are labelled with integers ≥ 3 or with the symbol ∞. If S denotes the set of vertices, let $m(s, s')$ denote the label on the edge joining $s \neq s'$. Since the label 3 occurs frequently, we omit it when drawing pictures. We also make the convention that $m(s, s') = 2$ for vertices $s \neq s'$ not joined by an edge, while $m(s, s) = 1$. (It will be seen in Chapter 5 that every Coxeter graph comes from some Coxeter group.)

We associate to a Coxeter graph Γ with vertex set S of cardinality n a symmetric $n \times n$ matrix A by setting

$$a(s, s') := -\cos \frac{\pi}{m(s, s')}.$$

Recall some terminology. Any symmetric $n \times n$ matrix $A = A^t$ defines a bilinear form $x^t A y \, (x, y \in \mathbf{R}^n)$, with associated quadratic form $x^t A x$. It is well known that the eigenvalues of A are all real. A is called **positive definite** if $x^t A x > 0$ for all $x \neq 0$, **positive semidefinite** if $x^t A x \geq 0$ for all x. Equivalently, A is positive definite if all its eigenvalues are (strictly) positive, positive semidefinite if all its eigenvalues are nonnegative. By abuse of language, we also say that A is of **positive type** if it is positive semidefinite, including positive definite. (This should not be confused with the notion of 'positive matrix', meaning one whose entries are strictly positive.) For brevity, we call Γ positive definite or positive semidefinite when the associated matrix (or bilinear form) has the corresponding property.

There is another well-known characterization of positive type in terms of determinants, which we shall use in the following two sections. The **principal minors** of A are the determinants of the submatrices obtained by removing the last k rows and columns ($0 \leq k < n$). Then A is positive definite (resp. positive semidefinite) if and only if all its principal minors are positive (resp. nonnegative).

When Γ comes from a finite reflection group W, the matrix A is in fact *positive definite*, because it represents the standard euclidean inner product relative to the basis Δ of V (assumed for convenience to consist of unit vectors). Our strategy for classifying finite reflection groups is to assemble a list of all possible connected positive definite Coxeter graphs, then to show that each of them does in fact correspond to a finite reflection group.

2.4   Some positive definite graphs

We claim that the graphs in Figure 1 are all positive definite. To verify

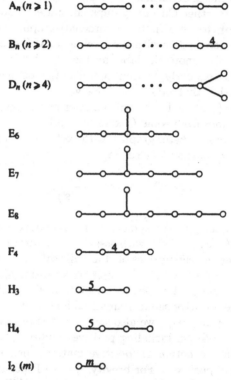

Figure 1: Some positive definite graphs

this we should compute the principal minors of the corresponding matrix A. It is clear by inspection that (with suitable numbering of vertices) each minor is itself the determinant of the matrix belonging to one of the graphs in Figure 1. So by induction on n (the number of vertices), it will be enough to compute det A itself in each case. Because the denominator 2 occurs so often, it is actually more convenient to compute det $2A$.

The cases $n \leq 2$ can be checked directly. For example, the matrix A corresponding to the graph $I_2(m)$ is

$$\begin{pmatrix} 1 & -\cos{(\pi/m)} \\ -\cos{(\pi/m)} & 1 \end{pmatrix}$$

Thus $\det 2A = 4(1 - \cos^2(\pi/m)) = 4\sin^2(\pi/m) > 0$.

If $n \geq 3$, a glance at Figure 1 shows that it is possible to number vertices in such a way that the last vertex (numbered n) is joined by an edge to only one other vertex (numbered $n-1$), this edge being labelled $m = 3$ or 4. Let d_i be the determinant of the upper left $i \times i$ submatrix of $2A$. Then an expansion of $\det 2A$ along the last row shows that

$$\det 2A = 2d_{n-1} - cd_{n-2}, \tag{1}$$

where $c = 1$ (resp. 2) if $m = 3$ (resp. 4). Keeping in mind that we multiply each matrix by 2, we compute inductively the values in the following table.

A_n	B_n	D_n	E_6	E_7	E_8	F_4	H_3	H_4	$I_2(m)$
$n+1$	2	4	3	2	1	1	$3-\sqrt{5}$	$(7-3\sqrt{5})/2$	$4\sin^2(\pi/m)$

Table 1: Determinant of $2A$

The reader should carry out the required verification as an exercise, recalling the values:

$$\cos\frac{\pi}{3} = \frac{1}{2}, \ \cos\frac{\pi}{4} = \frac{\sqrt{2}}{2}, \ \cos\frac{\pi}{5} = \frac{1+\sqrt{5}}{4}, \ \cos\frac{\pi}{6} = \frac{\sqrt{3}}{2}.$$

(If the value of $\cos \pi/5$ is not so familiar, look at the derivation in Bourbaki [1], p. 192 (footnote).)

As an example, we work through the cases of H_3 and H_4. For H_3 the smaller minors come from graphs of types $I_2(5)$ and A_1, so formula (1) reads:

$$\det 2A = 8\sin^2(\pi/5) - 2 = 3 - \sqrt{5}.$$

For H_4 the smaller minors are of types H_3 and $I_2(5)$, yielding:

$$\det 2A = 2(3 - \sqrt{5}) - 4\sin^2(\pi/5) = (7 - 3\sqrt{5})/2.$$

2.5 Some positive semidefinite graphs

As a tool in the proof that the Coxeter graphs in Figure 1 of 2.4 are the only connected positive definite ones, we assemble some auxiliary graphs in Figure 2. We claim that all of these are positive semidefinite (but not positive definite). The labels are suggestive of the fact that each graph is obtained from a graph in Figure 1 by adding a single vertex. In each case, the subscript n therefore indicates that the number of vertices is $n + 1$. (For type B_n there are two related graphs, labelled $\widetilde{B}_n$ and $\widetilde{C}_n$. We write G_2 in place of $I_2(6)$.) The actual significance of the graphs in Figure 2 will only become clear in Chapter 4.

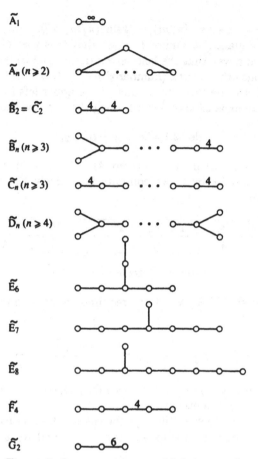

Figure 2: Some positive semidefinite graphs

Since the removal of a suitable vertex from each graph in Figure 2 leaves one of the positive definite graphs in Figure 1, all we have to check is that the determinant of the matrix A belonging to each graph is 0. This is immediately clear for type $\widetilde{A_n}$, since the sum of all rows in A is 0 and hence A is singular. For the remaining types we can use the inductive formula (1) and the table in 2.4. For example, consider $\widetilde{F_4}$. The relevant subgraphs are of types F_4 and B_3, so formula (1) reads:

$$\det 2A = 2 - 2 = 0.$$

It will be useful later on to know that the Coxeter graphs Z_4, Z_5 shown below are not of positive type. This follows from the fact that the determinant of $2A$ is (respectively) $3 - 2\sqrt{5}, 4 - 2\sqrt{5}$ (each of which is strictly negative). These are quickly computed via (1) in 2.4, using the determinants found there for types H_3 and H_4.

$$Z_4 \quad \circ\!\!-\!\!\circ \overset{5}{-\!\!\!-} \circ\!\!-\!\!\circ$$

$$Z_5 \quad \circ \overset{5}{-\!\!\!-} \circ\!\!-\!\!\circ -\!\!\!- \circ -\!\!\!- \circ$$

2.6 Subgraphs

Here we obtain a crucial fact for the classification program: each (proper) 'subgraph' of a connected graph of positive type is positive definite. By a **subgraph** of a Coxeter graph Γ we mean a graph Γ' obtained by omitting some vertices (and adjacent edges) or by decreasing the labels on one or more edges, or both. We also say that Γ 'contains' Γ'. To simplify statements, we choose not to call the graph itself a subgraph.

We shall need some standard (but possibly unfamiliar) facts from matrix theory, usually associated with the Perron–Frobenius theory of nonnegative matrices and M-matrices. The following general proposition will also play a key role in 3.17 and 6.5 below.

A real $n \times n$ matrix A is called **indecomposable** if there is no partition of the index set into nonempty subsets I, J such that $a_{ij} = 0$ whenever $i \in I, j \in J$. Otherwise, after renumbering indices, A could be written in block diagonal form. (The less exact term 'irreducible' is more commonly used in linear algebra texts.) It is clear that the matrix belonging to a Coxeter graph is indecomposable precisely when the graph is connected.

Proposition *Let A be a real symmetric $n \times n$ matrix which is positive semidefinite and indecomposable. (In particular, the eigenvalues of A are real and nonnegative.) Assume that $a_{ij} \leq 0$ whenever $i \neq j$. Then:*

(a) *$N := \{x \in \mathbf{R}^n | x^t A x = 0\}$ coincides with the nullspace of A and has dimension ≤ 1.*

(b) *The smallest eigenvalue of A has multiplicity 1, and has an eigenvector whose coordinates are all strictly positive.*

Proof. (a) It is clear that the nullspace of A lies in N. For the reverse inclusion, we diagonalize A. Since A is symmetric, there is an orthogonal matrix P for which $D := P^t A P = \mathrm{diag}(d_1, \ldots, d_n)$. If $0 = y^t D y = \sum d_i y_i^2$, then for each i either $d_i = 0$ or else $y_i = 0$ (since $d_i \geq 0$). Thus $\sum d_i y_i = 0$, and y lies in the nullspace of D. In turn, if $x = Py$ satisfies $x^t A x = 0$, we see that x lies in the nullspace of A.

Suppose N has positive dimension, say $0 \neq x \in N$. Let z be the vector whose coordinates are the absolute values of those of x. Since $a_{ij} \leq 0$ whenever $i \neq j$, we have

$$0 \leq z^t A z \leq x^t A x = 0,$$

forcing z to lie in N. We claim that all coordinates of z are nonzero. To see this, let J be the (nonempty) set of indices j for which $z_j \neq 0$, and let I be its complement. Since N is the nullspace of A, we have $\sum_j a_{ij} z_j = 0$ for each i, where the sum need only be taken over $j \in J$. Since $z_j > 0$ and $a_{ij} \leq 0$, each term in the sum is nonpositive. If I were nonempty we would get $a_{ij} = 0$ for all $i \in I, j \in J$, contrary to the indecomposability of A. Thus N contains a vector whose coordinates are all strictly positive. The argument also shows that an *arbitrary* nonzero element $x \in N$ has no zero coordinate. If dim N were larger than 1, it would be easy to find a nonzero linear combination of such vectors with a coordinate equal to 0, so we conclude that dim $N \leq 1$.

(b) Recalling that the eigenvalues d_i of A are nonnegative, let d be the smallest one. Observe that $A - dI$ satisfies all the hypotheses of the proposition. (It is positive semidefinite, since it is orthogonally similar to the matrix $D - dI$ with nonnegative entries.) Moreover, $A - dI$ is singular. So its nullspace has dimension exactly 1 and is spanned by a vector with strictly positive coefficients, according to the argument in (a). This means that d occurs as an eigenvalue of A with multiplicity 1, and there is a corresponding positive eigenvector. $\square$

Corollary *If Γ is a connected Coxeter graph of positive type, then every (proper) subgraph is positive definite.*

Proof. Let Γ' be a subgraph, and denote by A and A' the associated matrices, so that A' is $k \times k$ for some $k \leq n$. The edge labels in Γ' satisfy $m'_{ij} \leq m_{ij}$, whence $a'_{ij} = -\cos(\pi/m'_{ij}) \geq -\cos(\pi/m_{ij}) = a_{ij}$. Suppose A' fails to be positive definite. Then there is a nonzero vector $x \in \mathbf{R}^k$ such that $x^t A' x \leq 0$. Applying the quadratic form associated with A to the vector with coordinates $|x_1|, \ldots, |x_k|, 0, \ldots, 0$ in $\mathbf{R}^n$, we get the comparison:

$$0 \leq \sum a_{ij}|x_i||x_j| \leq \sum a'_{ij}|x_i||x_j| \leq \sum a'_{ij}x_i x_j \leq 0,$$

where each sum is taken over all $i, j \leq k$. (In the next-to-last inequality, we used the fact that $a'_{ij} \leq 0$ for $i \neq j$.) So equality holds throughout. The first equality shows that we have a null vector for A, which by the proposition is possible only if $k = n$ and all coordinates of x are nonzero. But then the second equality forces all $a_{ij} = a'_{ij}$, contrary to the assumption that Γ' is a (proper) subgraph. $\square$

2.7   Classification of graphs of positive type

Theorem *The graphs in Figure 1 of 2.4 and Figure 2 of 2.5 are the only connected Coxeter graphs of positive type.*

Proof. Suppose there were a connected Coxeter graph Γ of positive type not pictured in either Figure 1 or Figure 2. We proceed in 20 easy steps to obtain a contradiction, relying repeatedly on Corollary 2.6 to rule out various subgraphs. Let Γ have n vertices, and let m be the maximum edge label.

(1) All Coxeter graphs of rank 1 or 2 are clearly of positive type (A_1, $I_2(m)$, $\widetilde{A_1}$), so we must have $n \geq 3$.

(2) Since $\widetilde{A_1}$ cannot be a subgraph of Γ, we must have $m < \infty$.

(3) Since $\widetilde{A_n}(n \geq 2)$ cannot be a subgraph of Γ, Γ contains no circuits. Suppose for the moment that $m = 3$.

(4) Γ must have a branch point, since $\Gamma \neq A_n$.

(5) Γ contains no $\widetilde{D_n}, n > 4$, so it has a unique branch point.

(6) Γ does not contain $\widetilde{D_4}$, so exactly three edges meet at the branch point (with $a \leq b \leq c$ further vertices lying in these three directions).

(7) Since $\widetilde{E_6}$ is not a subgraph of Γ, $a = 1$.

(8) Since $\widetilde{E_7}$ is not a subgraph of Γ, $b \leq 2$.

(9) Since $\Gamma \neq D_n$, b cannot be 1, so $b = 2$.

(10) Since $\widetilde{E_8}$ is not a subgraph of Γ, $c \leq 4$.

(11) Since $\Gamma \neq E_6, E_7, E_8$, the case $m = 3$ is impossible. Thus $m \geq 4$.

(12) Γ does not contain $\widetilde{C_n}$, so only one edge has a label > 3.

(13) Γ does not contain $\widetilde{B_n}$, so Γ has no branch point. Now consider what happens if $m = 4$.

(14) Since $\Gamma \neq B_n$, the two extreme edges of Γ are labelled 3.

(15) Since Γ does not contain $\widetilde{F_4}$, n must be 4.

(16) But $\Gamma \neq F_4$, so the case $m = 4$ is impossible. Thus $m \geq 5$.

(17) Since Γ does not contain $\widetilde{G_2}$, we must have $m = 5$.

(18) Γ does not contain the nonpositive graph Z_4 in 2.5, so the edge labelled 5 must be an extreme edge.

(19) Γ does not contain the nonpositive graph Z_5, so $n \leq 4$.

(20) Now Γ must be either H_3 or H_4, which is absurd. So we have eliminated all possibilities. $\square$

The theorem limits the possible finite reflection groups which can exist. In fact, there does exist a group belonging to each graph in Figure 1 of 2.4. A uniform existence proof will be given in Chapters 5 and 6, along the following lines. Define W abstractly by generators and relations, as in Theorem 1.9. (It will not yet be obvious whether W is finite or not.) Then show how to represent W faithfully as a subgroup generated by reflections in a suitable $GL(V)$, and argue that W is finite (being a discrete subgroup of the compact group $O(V)$).

In the rest of this chapter we shall discuss concretely how to construct finite reflection groups of all types, and thereby determine their orders.

2.8   Crystallographic groups

A subgroup G of $GL(V)$ is said to be **crystallographic** if it stabilizes a lattice L in V (the $\mathbf{Z}$-span of a basis of V): $gL \subset L$ for all $g \in G$. (Since G is a group, it is automatic that $gL = L$.) The name comes from low-dimensional crystallography, where the classification of possible crystal structures depends heavily on the available symmetry groups. It turns out that 'most' finite reflection groups are crystallographic.

First we obtain a necessary condition for W to be crystallographic. The crucial thing to notice is that, for any choice of basis in V, the traces of the matrices representing a crystallographic group must be in $\mathbf{Z}$ (since the trace could equally well be computed relative to a $\mathbf{Z}$-basis of L).

Proposition *If W is crystallographic, then each integer $m(\alpha, \beta)$ must be $2, 3, 4$, or 6 when $\alpha \neq \beta$ in Δ.*

Proof. If $\alpha \neq \beta$, we know that $s_\alpha s_\beta \neq 1$ acts on the plane spanned by α and β as a rotation through the angle $\theta := 2\pi/m(\alpha, \beta)$, while fixing the orthogonal complement pointwise. Thus its trace relative to a compatible choice of basis for V is $(n - 2) + 2\cos\theta$ $(n = \dim V)$. So $\cos\theta$ must be a half-integer, while $0 < \theta \leq \pi$. The only possibilities are $\cos\theta = -1, -1/2, 0, 1/2$, corresponding to the cases $m(\alpha, \beta) = 2, 3, 4, 6$. □

This criterion rules out the groups of types H_3 and H_4 as well as all dihedral groups except those of orders 2, 4, 6, 8, 12. For all remaining cases, we shall see in the following sections a 'natural' construction of a W-stable lattice. But it is easy enough to show in an *ad hoc* way that such lattices exist:

Exercise. If W satisfies the necessary condition in the proposition (and is essential), show how to modify the lengths of the roots in a simple system Δ so that the resulting $\mathbf{Z}$-span is a lattice in V stable under W. [Use the fact that the Coxeter graph has no circuits.]

2.9 Crystallographic root systems and Weyl groups

The notion of 'root system' introduced in 1.2 differs somewhat from that commonly used in Lie theory. To avoid confusion, we say that a root system Φ is **crystallographic** if it satisfies the additional requirement:

$$(R3) \qquad \frac{2(\alpha,\beta)}{(\beta,\beta)} \in \mathbf{Z} \text{ for all } \alpha,\beta \in \Phi.$$

These integers are called **Cartan integers**. It is actually enough to require that the ratios be integers when $\alpha,\beta \in \Delta$. The group W generated by all reflections s_α ($\alpha \in \Phi$) is known as the **Weyl group** of Φ.

The effect of the added condition on Φ is to insure that $s_\alpha\beta$ is obtained from β by adding an *integral* multiple of α. This in turn implies that all roots are $\mathbf{Z}$-linear combinations of Δ, and that (in case W is essential) the $\mathbf{Z}$-span of Δ in V is a W-stable lattice. So W is crystallographic in the sense of 2.8.

We shall not give details of the classification (up to 'isomorphism') of crystallographic root systems, since it is similar in spirit to the classification of positive definite Coxeter graphs given earlier. (See Chapter VI of Bourbaki [1] or Chapter III of Humphreys [1].) The conclusion is that the resulting Weyl groups are precisely the reflection groups for which all $m(\alpha,\beta) \in \{2,3,4,6\}$ (when $\alpha \neq \beta$). So Weyl groups are the same thing as crystallographic reflection groups. However, there are distinct crystallographic root systems B_n and C_n, each having as Weyl group the group previously labelled B_n.

It turns out that when Φ (or W) is irreducible (in the sense of 2.2), at most two root lengths are possible. If there are both 'long' and 'short' roots, the ratio of the squared lengths can only be 2 or 3. (If there is just one root length, all roots are called 'long'.) This information is added to the Coxeter graph by directing an arrow toward the short root when adjacent vertices represent a long and a short root. By convention, the label 4 or 6 is replaced in each case by a double or triple edge. (When only one root length occurs, the graph is therefore 'simply-laced'.) The resulting **Dynkin diagrams** are easily derived from Figure 1 in 2.4 together with 2.10 below. (See Bourbaki [1], p. 197 or Humphreys [1], p. 58.)

The construction of the various crystallographic root systems will be outlined in the following section, following Bourbaki. For later reference, we summarize a few general facts:

(1) Setting $\alpha^\vee := 2\alpha/(\alpha,\alpha)$, the set $\Phi^\vee$ of all **coroots** $\alpha^\vee$ ($\alpha \in \Phi$) is also a crystallographic root system in V, with simple system $\Delta^\vee := \{\alpha^\vee | \alpha \in \Delta\}$. It is called the **inverse** or **dual** root system. The Weyl group of $\Phi^\vee$ is W, with $w\alpha^\vee = w(\alpha)^\vee$. In most cases $\Phi^\vee$ is isomorphic to

Φ; however, the root systems of types B_n and C_n are dual to each other. Short roots α in a system Φ of type B_n give rise to long roots $\alpha^\vee$ in the system $\Phi^\vee$ of type C_n (and vice versa). (*Note:* Rather than working in a euclidean space, Bourbaki defines root systems in an arbitrary real vector space, with coroots belonging to the dual vector space. This does not significantly alter any of the results we are quoting.)

(2) The $\mathbf{Z}$-span $L(\Phi)$ of Φ in V is called the **root lattice**; it is a lattice in the subspace of V spanned by Φ, which we can usually assume to be V itself. Similarly, we define the **coroot lattice** $L(\Phi^\vee)$. Both lattices are W-stable. In Chapter 4 it will also be important to introduce related W-stable lattices (which arise in representation theory). Define the **weight lattice**

$$\hat{L}(\Phi) := \{\lambda \in V | (\lambda, \alpha^\vee) \in \mathbf{Z} \text{ for all } \alpha \in \Phi\},$$

and the **coweight lattice**

$$\hat{L}(\Phi^\vee) := \{\lambda \in V | (\lambda, \alpha) \in \mathbf{Z} \text{ for all } \alpha \in \Phi\}.$$

Then $\hat{L}(\Phi)$ contains $L(\Phi)$ as a subgroup of finite index f, and similarly $\hat{L}(\Phi^\vee)$ contains $L(\Phi^\vee)$ as a subgroup of index f. Here f is just the determinant of the matrix of Cartan integers $(\alpha, \beta^\vee)$ $(\alpha, \beta \in \Delta)$. (It is called the **index of connection** in Lie theory: $\hat{L}/L$ is isomorphic to the fundamental group of a compact Lie group of adjoint type having W as Weyl group; so f is the order of the kernel of the associated map from the simply connected covering group.)

(3) There is a natural partial ordering on V (when Δ is fixed): $\mu \le \lambda$ if and only if $\lambda - \mu$ is a nonnegative $\mathbf{Z}$-linear combination of Δ. When Φ is irreducible, there exists a unique highest root (a long root) relative to this ordering, denoted $\tilde{\alpha}$; it plays a crucial role in 2.11 below as well as in Chapter 4. There also exists a unique highest short root. (This is easy to prove in the axiomatic framework of root systems, but is most easily understood in terms of the adjoint representation of the simple Lie algebra over $\mathbf{C}$ having Φ as root system.)

(4) The long (resp. short) roots form a single orbit under the permutation action of W on Φ, assuming W is irreducible. This is seen as follows if all roots are long. By Corollary 1.5, each root is W-conjugate to a simple root, and since the Dynkin diagram is connected, it then suffices to see that roots belonging to adjacent vertices are W-conjugate. But they are clearly in the same orbit under the subgroup (isomorphic to $\mathcal{D}_3$) generated by the two simple reflections involved. (More generally, it is easy to check that all reflections are conjugate in $\mathcal{D}_m$ whenever m is odd.) If there are two root lengths, the argument is similar, since simple roots of each length correspond to a connected part of the Dynkin diagram in types B_n, C_n, F_4, G_2.

Remark. In Lie theory, one sometimes wants to allow both α and 2α to be 'roots'. If the condition (R1) in 1.2 is dropped from the definition of crystallographic root system, Φ may be 'nonreduced' (as permitted in Bourbaki's definition of 'root system'). The only new (irreducible) type one gets is called BC_n: superimpose root systems of type B_n and C_n, the long roots of B_n coinciding with the short roots of C_n.

2.10   Construction of root systems

Denote by $\varepsilon_1, \ldots, \varepsilon_n$ the standard basis of $\mathbf{R}^n$. Whenever we write combinations such as $\pm\varepsilon_i \pm \varepsilon_j$ below, it is understood that the signs may be chosen arbitrarily.

We shall outline briefly the construction of (crystallographic) root systems of all possible types, following Chapter VI of Bourbaki [1] (which should be consulted for further details). The basic strategy is simple enough. In a suitably chosen lattice L in $\mathbf{R}^n$, define Φ to be the set of all vectors having one or two prescribed lengths. Then check that the resulting scalars $2(\alpha, \beta)/(\beta, \beta)$ lie in $\mathbf{Z}$. It follows automatically that the reflections with respect to vectors in Φ stabilize L and hence permute Φ as required. (The actual choices to be made are not so obvious; they arose historically from a close scrutiny of the adjoint representation of a simple Lie algebra.) The reader should be able to fill in the calculations without difficulty.

In effect we already encountered root systems of types A_n, B_n, D_n in 1.1 (and C_n is just the dual of B_n). In these cases W has a fairly simply description. In the 'exceptional' cases E_6, E_7, E_8, F_4, G_2, it is harder to deduce from the description of Φ a concrete description of W (apart from the easy case of G_2). In particular, the order of W remains to be calculated. We shall develop a general method for this in 2.11 below, and in 2.12 we shall describe some realizations of the exceptional Weyl groups.

$(A_n, n \geq 1)$ Let V be the hyperplane in $\mathbf{R}^{n+1}$ consisting of vectors whose coordinates add up to 0. Define Φ to be the set of all vectors of squared length 2 in the intersection of V with the standard lattice $\mathbf{Z}\varepsilon_1 + \ldots + \mathbf{Z}\varepsilon_{n+1}$. Then Φ consists of the $n(n+1)$ vectors:

$$\varepsilon_i - \varepsilon_j \ (1 \leq i \neq j \leq n+1).$$

For Δ take

$$\alpha_1 = \varepsilon_1 - \varepsilon_2, \ \alpha_2 = \varepsilon_2 - \varepsilon_3, \ \ldots, \ \alpha_n = \varepsilon_n - \varepsilon_{n+1}.$$

Then $\tilde{\alpha} = \varepsilon_1 - \varepsilon_{n+1}$. W is S_{n+1}, which acts as usual by permuting the ε_i.

($B_n, n \geq 2$) Let $V = \mathbf{R}^n$, and define Φ to be the set of all vectors of squared length 1 or 2 in the standard lattice. So Φ consists of the $2n$ short roots $\pm\varepsilon_i$ and the $2n(n-1)$ long roots $\pm\varepsilon_i\pm\varepsilon_j(i < j)$, totalling $2n^2$. For Δ take $\alpha_1 = \varepsilon_1 - \varepsilon_2$, $\alpha_2 = \varepsilon_2 - \varepsilon_3, \ldots, \alpha_{n-1} = \varepsilon_{n-1} - \varepsilon_n$, $\alpha_n = \varepsilon_n$. Then $\tilde{\alpha} = \varepsilon_1 + \varepsilon_2$. W is the semidirect product of S_n (which permutes the ε_i) and $(\mathbf{Z}/2\mathbf{Z})^n$ (acting by sign changes on the ε_i), the latter normal in W.

($C_n, n \geq 2$) Starting with B_n, one can define C_n to be its inverse root system. (Note that B_2 and C_2 are isomorphic.) It consists of the $2n$ long roots $\pm 2\varepsilon_i$ and the $2n(n-1)$ short roots $\pm\varepsilon_i \pm \varepsilon_j$ $(i < j)$. For Δ take $\alpha_1 = \varepsilon_1 - \varepsilon_2$, $\alpha_2 = \varepsilon_2 - \varepsilon_3, \ldots, \alpha_{n-1} = \varepsilon_{n-1} - \varepsilon_n$, $\alpha_n = 2\varepsilon_n$. Then $\tilde{\alpha} = 2\varepsilon_1$.

($D_n, n \geq 4$) Let $V = \mathbf{R}^n$, and define Φ to be the set of all vectors of squared length 2 in the standard lattice. So Φ consists of the $2n(n-1)$ roots $\pm\varepsilon_i \pm \varepsilon_j$ $(1 \leq i < j \leq n)$. For Δ take $\alpha_1 = \varepsilon_1 - \varepsilon_2$, $\alpha_2 = \varepsilon_2 - \varepsilon_3, \ldots, \alpha_{n-1} = \varepsilon_{n-1} - \varepsilon_n$, $\alpha_n = \varepsilon_{n-1} + \varepsilon_n$. Then $\tilde{\alpha} = \varepsilon_1 + \varepsilon_2$. W is the semidirect product of S_n (permuting the ε_i) and $(\mathbf{Z}/2\mathbf{Z})^{n-1}$ (acting by an even number of sign changes), the latter normal in W.

(G_2) Let V be the hyperplane in $\mathbf{R}^3$ consisting of vectors whose coordinates add up to 0. Define Φ to be the set of vectors of squared length 2 or 6 in the intersection of V with the standard lattice. So Φ consists of six short roots $\pm(\varepsilon_i - \varepsilon_j)$ $(i < j)$ and six long roots $\pm(2\varepsilon_i - \varepsilon_j - \varepsilon_k)$ (where $\{i,j,k\} = \{1,2,3\}$). For Δ take $\alpha_1 = \varepsilon_1 - \varepsilon_2$, $\alpha_2 = -2\varepsilon_1 + \varepsilon_2 + \varepsilon_3$. Then $\tilde{\alpha} = 2\varepsilon_3 - \varepsilon_1 - \varepsilon_2$.

(F_4) Let $V = \mathbf{R}^4$. If L' is the standard lattice, let $L := L' + \mathbf{Z}(\frac{1}{2}\sum_{i=1}^{4}\varepsilon_i)$. This is also a lattice, and we define Φ to be the set of all vectors in L of squared length 1 or 2. So Φ consists of 24 long roots and 24 short roots:

$$\pm\varepsilon_i \pm \varepsilon_j \ (i < j),$$

$$\pm\varepsilon_i, \ \frac{1}{2}(\pm\varepsilon_1 \pm \varepsilon_2 \pm \varepsilon_3 \pm \varepsilon_4).$$

For Δ take

$$\alpha_1 = \varepsilon_2 - \varepsilon_3, \ \alpha_2 = \varepsilon_3 - \varepsilon_4, \ \alpha_3 = \varepsilon_4, \ \alpha_4 = \frac{1}{2}(\varepsilon_1 - \varepsilon_2 - \varepsilon_3 - \varepsilon_4).$$

Then $\tilde{\alpha} = \varepsilon_1 + \varepsilon_2$.

Since a root system of type E_8 must contain canonical copies of both E_7 and E_6, the main task is to construct the former.

(E_8) Let $V = \mathbf{R}^8$. The choice of lattice is somewhat subtle. Start with the lattice L' consisting of all $\sum c_i\varepsilon_i$ with $c \in \mathbf{Z}$ and $\sum c_i$ even.

Then let $L = L' + \mathbf{Z}(\frac{1}{2}\sum_{i=1}^{8} \varepsilon_i)$. Define Φ to be the set of all vectors of squared length 2 in L. So Φ consists of 240 roots:

$$\pm\varepsilon_i \pm \varepsilon_j \ (i < j), \quad \frac{1}{2}\sum_{i=1}^{8} \pm\varepsilon_i \ \text{(even number of + signs).}$$

For Δ take

$$\begin{aligned}
\alpha_1 &= \frac{1}{2}(\varepsilon_1 - \varepsilon_2 - \varepsilon_3 - \varepsilon_4 - \varepsilon_5 - \varepsilon_6 - \varepsilon_7 + \varepsilon_8), \\
\alpha_2 &= \varepsilon_1 + \varepsilon_2, \\
\alpha_i &= \varepsilon_{i-1} - \varepsilon_{i-2} \ (3 \le i \le 8).
\end{aligned}$$

Then $\tilde{\alpha} = \varepsilon_7 + \varepsilon_8$.

(E_7) Starting with the root system of type E_8 just constructed, let V be the span of the α_i ($1 \le i \le 7$) in $\mathbf{R}^8$, and let Φ be the set of 126 roots of E_8 lying in V:

$$\pm\varepsilon_i \pm \varepsilon_j \ (1 \le i < j \le 6),$$

$$\pm(\varepsilon_7 - \varepsilon_8),$$

$$\pm\frac{1}{2}(\varepsilon_7 - \varepsilon_8 + \sum_{i=1}^{6} \pm\varepsilon_i),$$

where the number of minus signs in the sum is *odd*. The roots α_i ($1 \le i \le 7$) form a simple system. Then $\tilde{\alpha} = \varepsilon_8 - \varepsilon_7$.

(E_6) Start again with the root system of type E_8, and let V be the span of the α_i ($1 \le i \le 6$), with Φ defined to be the set of 72 roots of E_8 lying in V:

$$\pm\varepsilon_i \pm \varepsilon_j \ (1 \le i < j \le 5),$$

$$\pm\frac{1}{2}(\varepsilon_8 - \varepsilon_7 - \varepsilon_6 + \sum_{i=1}^{5} \pm\varepsilon_i),$$

where the number of minus signs in the sum is *odd*. The roots α_i ($1 \le i \le 6$) form a simple system. Then $\tilde{\alpha} = \frac{1}{2}(\varepsilon_1 + \varepsilon_2 + \varepsilon_3 + \varepsilon_4 + \varepsilon_5 - \varepsilon_6 - \varepsilon_7 + \varepsilon_8)$.

Exercise. In each case, express $\tilde{\alpha}$ as a $\mathbf{Z}$-linear combination of the simple roots.

2.11 Computing the order of W

In cases where we have a 'natural' construction of W (types A_n, B_n, D_n, $I_2(m)$), there is no problem about computing $|W|$. But in other cases the order is not immediately apparent even after Φ is exhibited. Here we describe a general method, based on the elementary group-theoretic fact: if a finite group G acts as a permutation group on a set X then, for each $x \in X$, $|G| = |Gx||G_x|$, where Gx is the orbit of x and G_x is the isotropy group of x in G. In our situation, W acts on the set Φ, and for a well-chosen root we can describe the isotropy group explicitly. Here we consider just Weyl groups, deferring discussion of H_3 and H_4 to 2.13.

The key fact, stated as (4) in 2.9, is that all long (resp. short) roots form a single W-orbit (when W is irreducible). Consider the unique highest root $\tilde{\alpha}$ (which is long). If $\alpha \in \Delta$, we claim that $(\tilde{\alpha}, \alpha) \geq 0$. Otherwise $s_\alpha \tilde{\alpha}$ would equal $\tilde{\alpha}$ plus a positive multiple of α, hence would be higher in the partial ordering. Thus $\tilde{\alpha}$ lies in the fundamental domain D for W described in Theorem 1.12. According to part (a) of that theorem, the isotropy group of $\tilde{\alpha}$ is generated by the reflections belonging to *simple* roots orthogonal to $\tilde{\alpha}$. These are easily determined in each case from the data assembled in 2.10. This gives an inductive method for calculating $|W|$, which we apply now to the remaining types.

(F_4) There are 24 long roots. The highest root $\tilde{\alpha}$ is orthogonal to all simple roots except α_1, so its isotropy group is of type C_3 (having order 48). Therefore $|W| = 24 \cdot 48 = 1152 = 2^7 3^2$.

(E_6) The 72 roots form a single orbit. The highest root $\tilde{\alpha}$ is orthogonal to all simple roots except α_2, so its isotropy group is of type A_5 (having order 6!). Therefore $|W| = 6!72 = 2^7 3^4 5$.

(E_7) The 126 roots form a single orbit. The highest root $\tilde{\alpha}$ is orthogonal to all simple roots except α_1, so its isotropy group is of type D_6 (having order $2^5 6!$). Therefore $|W| = 2^5 6!126 = 2^{10} 3^4 5 \, 7$.

(E_8) The 240 roots form a single orbit. The highest root $\tilde{\alpha}$ is orthogonal to all simple roots except α_8, so its isotropy group is of type E_7 (having the order just computed). Therefore $|W| = 2^{14} 3^5 5^2 7$.

To summarize this discussion, the following table gives the orders of all irreducible Weyl groups (along with the number of roots):

A_n	B_n/C_n	D_n	E_6	E_7	E_8	F_4	G_2
$(n+1)!$	$2^n n!$	$2^{n-1} n!$	$2^7 3^4 5$	$2^{10} 3^4 5 \, 7$	$2^{14} 3^5 5^2 7$	$2^7 3^2$	12
$n(n+1)$	$2n^2$	$2n(n-1)$	72	126	240	48	12

Table 2: $|W|$ and $|\Phi|$ for Weyl groups

Some other formulas for $|W|$ will be developed in 3.9 and 4.9.

Exercise. Use the method of this section to derive again the orders of the groups of types A_n, B_n, D_n.

2.12 Exceptional Weyl groups

Here we survey briefly some of the interesting ways in which the Weyl groups of types F_4, E_6, E_7, E_8 can be described. Only F_4 arises as the group of symmetries of a regular solid, but the other three groups have as close relatives certain simple groups: orthogonal and symplectic groups over F_2 or F_3. (See the exercises in Bourbaki [1], pp. 228–229, as well as the references given below to the Atlas, Conway *et al.* [1].)

Recall that W always has a normal subgroup W^+ of index 2 (the 'rotation subgroup' consisting of elements of determinant 1).

(F_4) This is the group of symmetries of a regular solid in R^4 having 24 (three-dimensional) faces which are octahedra; see Coxeter [1]. Readers who share the author's inability to visualize such things may also welcome a purely group-theoretic description of W (see Bourbaki [1], p. 213). Observe that the 24 long roots in Φ form a root system Φ' of type D_4. It turns out that W is precisely the automorphism group of Φ', in which the Weyl group W' of type D_4 is a normal subgroup of order 192. (Recall that W' is the semidirect product of S_4 and an elementary abelian group of order 8.) Other automorphisms of Φ' arise naturally from symmetries of the Dynkin diagram: one can interchange the three outer vertices using S_3. So $W/W' \cong S_3$ (in agreement with the calculation $|W| = 2^7 3^2$ in 2.11 above). In fact, W is the semidirect product of W' and S_3.

(E_6) W has a number of interesting realizations (see the discussion in the Atlas under $U_4(2) \cong S_4(3)$). For example, W is the group of automorphisms of the famous configuration of 27 lines on a cubic surface. The rotation subgroup W^+ of W is a simple group of order 25 920, which has a variety of descriptions as a group of Lie type in the Atlas: $SU_4(2)$, $PSp_4(3)$, $SO_5(3)$, $O_6^-(2)$. One way to make such identifications is to pass to quotients of the root lattice. For example, $L(\Phi)/2L(\Phi)$ is a six-dimensional vector space over F_2. The usual inner product (divided by 2) induces a nondegenerate quadratic form on this space, invariant under the induced action of W. This yields an isomorphism of W onto the orthogonal group of the form.

(E_7) The rotation subgroup W^+ of W has two realizations as a simple group of Lie type, denoted $S_6(2)$ in the Atlas. On the one hand, $L(\Phi)/2L(\Phi)$ is a seven-dimensional vector space over F_2. The usual inner product (divided by 2) induces a nondegenerate quadratic form on this space, invariant under the induced action of W. This yields a homomorphism of W *onto* the orthogonal group of the form, with kernel $\{\pm 1\}$, inducing an isomorphism of W^+ onto $O_7(2)$. (This is a simple group of Lie type $B_3(2)$.) On the other hand, $L(\Phi)/2\hat{L}(\Phi)$ is a six-dimensional vector space over F_2, on which the usual inner product induces a nondegenerate alternating form with associated symplectic group $Sp_6(2)$ (a simple group of Lie type $C_3(2)$). Again W^+ is identified with this group.

(E_8) $L(\Phi)/2L(\Phi)$ is an eight-dimensional vector space over $\mathbf{F}_2$. The usual inner product (divided by 2) induces a nondegenerate quadratic form on this space, invariant under the induced action of W. This yields a homomorphism of W *onto* the orthogonal group of the form, with kernel $\{\pm 1\}$. The rotation subgroup W^+ of W maps onto a *simple* subgroup of index 2 in the orthogonal group, denoted $O_8^+(2)$ in the Atlas. This is a simple group of Lie type $D_4(2)$ having order $2^{12}3^5 5^2 7$.

2.13   Groups of types H_3 and H_4

Finally, we consider the non-crystallographic groups of types H_3 and H_4. Both of these arise naturally as symmetry groups of regular solids. The group of type H_3 is the symmetry group of the *icosahedron* (with 20 triangular faces) in $\mathbf{R}^3$, or dually, of the *dodecahedron* (with 12 pentagonal faces). W has order 120, contains 15 reflections, and is abstractly isomorphic to the direct product of its center $\{\pm 1\}$ and the simple group of order 60. (See Grove–Benson [1] for a detailed discussion of the 'icosahedral' group.) The group of type H_4 is the symmetry group of a regular 120-sided solid (with dodecahedral faces) in $\mathbf{R}^4$, or dually, of a regular 600-sided solid (with tetrahedral faces); see Coxeter [1], p. 153. It has order 14 400 and contains 60 reflections.

Rather than attempt a geometric construction of either group, we look for a suitable root system in $\mathbf{R}^3$ or $\mathbf{R}^4$. The existence of a group of type H_4 will of course imply the existence of a subgroup of type H_3, so we concentrate on the former. However, we can see in advance that the order of a group of type H_3 (if it exists) must be 120, by a simple application of the alternating sum formula derived in 1.11:

$$\sum_{I \subset S} (-1)^I \frac{|W|}{|W_I|} = 1.$$

Unfortunately, this formula is effective only for groups of odd rank, so the order of a group of type H_4 is not *a priori* obvious.

Probably the most insightful way to construct the root system of H_4 in $\mathbf{R}^4$ is to identify vectors with elements of the ring $\mathbf{H}$ of real quaternions: $\lambda = (c_1, c_2, c_3, c_4)$ corresponds to $\lambda = c_1 + c_2 i + c_3 j + c_4 k$ (where $\{1, i, j, k\}$ is the usual basis of $\mathbf{H}$). Under this identification the inner product (λ, μ) in $\mathbf{R}^4$ becomes:

$$\frac{1}{2}(\lambda \bar{\mu} + \mu \bar{\lambda}),$$

where $\bar{\lambda} := c_1 - c_2 i - c_3 j - c_4 k$ is the usual conjugation. Using the resulting norm $\|\lambda\| := \lambda \bar{\lambda}$, inverses in the division ring $\mathbf{H}$ are computed by

$$\lambda^{-1} = \frac{\bar{\lambda}}{\|\lambda\|}.$$

If $\alpha \in \mathbf{H}$ has norm 1, a quick calculation shows that the reflection s_α transforms $\mathbf{H}$ by the rule

$$\lambda \mapsto -\alpha\bar{\lambda}\alpha.$$

How can we locate the root system of H_4 as a subset of $\mathbf{H}$? The following (somewhat surprising) observation provides a clue:

Lemma *Any finite subgroup G of even order in $\mathbf{H}$ is a root system (when regarded as a subset of $\mathbf{R}^4$).*

Proof. Note first that each element of G must be of norm 1, since $\lambda^r = 1$ implies $\|\lambda\|^r = 1$ and hence $\|\lambda\| = 1$ (being a positive real number). In turn, since G is closed under inverses, it is closed under conjugation.

It is easy to check that no quaternion except -1 has multiplicative order 2. Since any group of even order contains an element of order 2, -1 must lie in G (and thus G contains the negatives of its elements). In turn, the formula for s_α shows that $s_\alpha G = G$ (G being closed under conjugation). Thus G satisfies the axioms of 1.2 for a root system. $\square$

For a systematic discussion of finite multiplicative subgroups of $\mathbf{H}$, consult DuVal [1], §20. Here we just specify a particular subgroup of order 120, without attempting to motivate the choice. It can be seen directly (or indirectly, using the classification) to be a root system of type H_4. This will prove the existence of a reflection group of type H_4 (and with it a subgroup of type H_3).

As the Coxeter graph suggests, the angle $\pi/5$ should figure in the construction. Set

$$a := \cos\frac{\pi}{5} = \frac{1+\sqrt{5}}{4}, \quad b := \cos\frac{2\pi}{5} = \frac{-1+\sqrt{5}}{4}.$$

Then one checks that

$$2a = 2b + 1, \quad 4ab = 1, \quad 4a^2 = 2a + 1, \quad 4b^2 = -2b + 1.$$

Let Φ consist of the unit vectors in $\mathbf{H}$ obtained from 1, $\frac{1}{2}(1 + i + j + k)$, and $a + \frac{1}{2}i + bj$ by even permutations of coordinates and arbitrary sign changes. It is routine (but tedious) to check that Φ is a group of order 120, hence is a root system. Let W be the resulting reflection group. It is easy to check that W is irreducible: it is impossible to partition Φ into two nonempty orthogonal subsets. From the classification it follows that W must be of type H_4. We can exhibit a simple system:

$$\alpha_1 = a - \frac{1}{2}i + bj$$

$$\alpha_2 \;=\; -a + \frac{1}{2}i + bj$$

$$\alpha_3 \;=\; \frac{1}{2} + bi - aj$$

$$\alpha_4 \;=\; -\frac{1}{2} - ai + bk$$

The resulting inner products are obviously consistent with the graph of type H_4. Rather than check directly that this is a simple system, the reader might consult Exercise 1 in 1.5.

To compute the order of W, we use the method of 2.11. Observe first that the 120 roots form a single W-orbit. Since every root is W-conjugate to a simple root, it is enough to check that simple roots belonging to adjacent vertices of the Coxeter graph are W-conjugate. Whether the edge is labelled 3 or 5, we can appeal as before to the general fact that all roots of a dihedral group $\mathcal{D}_m$ are in a single orbit when m is odd.

Note that the roots $\alpha_1, \alpha_2, \alpha_3$ form a simple system for the subgroup of type H_3, whose 30 roots are precisely the roots orthogonal to the root $k = (0,0,0,1)$. By part (c) of Theorem 1.12, the isotropy group of this root has order 120. Finally, $|W| = 120 \cdot 120 = 14\,400$.

Notes

In 2.3–2.7 we follow Witt [1]. (See also Chapter XI of Coxeter [1].)

(2.12) Coxeter [6] has more discussion of the groups of type E_n.

(2.13) For another description of the group of type H_4, exhibiting its structure, see Huppert [1]. We have mainly followed Witt [1]; note that the description in Grove–Benson [1] is slightly different. Sekiguchi–Yano [2] show how to embed H_3 in D_3; similarly, Shcherbak [1] embeds H_4 in E_8.

Chapter 3

Polynomial invariants of finite reflection groups

If W is a finite subgroup of $GL(V)$ generated by reflections, it acts in a natural way on the ring of polynomial functions on V. This chapter will be devoted to the study of this action, emphasizing the remarkable features of the subring of invariants, which turns out to be a polynomial ring on generators of certain well-determined degrees (whose product is $|W|$). This is a far-reaching generalization of the fundamental theorem on symmetric polynomials (the case of a symmetric group).

After some generalities on invariants of arbitrary finite groups (3.1)–(3.2), we prove the fundamental theorem of Chevalley giving an algebraically independent set of generators for the ring of invariants (3.3)–(3.5) and observe (3.7) that their degrees are uniquely defined. Moreover, the sum and product of the degrees have natural interpretations (3.9). A standard Jacobian criterion for algebraic independence of polynomials (3.10) allows us to work out some examples (3.12). The degrees enter in a surprising way into the factorization of the Poincaré polynomial of W (3.15).

In 3.16–3.19 we find a completely different interpretation of the degrees, in terms of the eigenvalues of a 'Coxeter element' of W (the product of simple reflections in some order). For Weyl groups the calculation of degrees can also be done by counting roots of each height (3.20).

3.1 Polynomial invariants of a finite group

Before dealing specifically with reflection groups, let us consider what can be said about the polynomial invariants of an arbitrary finite subgroup of $GL(V)$, where V is an n-dimensional vector space over a field K of characteristic 0. Denote by S the symmetric algebra $S(V^*)$ of

49

the dual space V^*, which is the algebra of polynomial functions on V. Relative to a fixed basis of V, S may be identified with the polynomial ring $K[x_1, \ldots, x_n]$, where the x_i are the coordinate functions. When no confusion can result, we sometimes write $K[x]$, $f(x)$, etc., for short. (The letter S has been used in Chapters 1 and 2 to denote a set of simple reflections generating W, and will be used again for that purpose later in this chapter. Meanwhile, the use of S to denote the symmetric algebra should cause no confusion.)

There is a natural action of G on S as a group of K-algebra automorphisms, coming from the contragredient action of G on V^*: $(g \cdot f)(v) = f(g^{-1}v)$, where $g \in G$, $v \in V$, $f \in V^*$. This action preserves the natural grading of S by 'degree'. We adopt the usual conventions that deg $0 = -\infty$ and that deg f is the maximum degree of the homogeneous parts of f. We say that $f \in S$ is G-**invariant** if $g \cdot f = f$ for all $g \in G$. Denote by $R = S^G$ the subalgebra of G-invariants. Note that it is homogeneous relative to the grading of S.

We want to get some feeling for the nature of R — for example, how 'big' is it? (The only obvious invariants are the constants.) It is instructive to compare the induced action of G on the field of fractions L of S, which is isomorphic to $K(x_1, \ldots, x_n)$, a purely transcendental extension of K of transcendence degree n. Here G acts as a group of field automorphisms. From field theory we know that L is a finite Galois extension of the fixed field L^G, with Galois group G. It follows that L^G also has transcendence degree n over K.

But how is L^G related to R? Obviously the field of fractions of R is included in L^G. We claim that the reverse inclusion is also true: Suppose $p/q \in L^G$ ($p, q \in S$). Both numerator and denominator may be multiplied by $\prod g \cdot p$, where the product is taken over all $g \neq 1$ in G. The new numerator is visibly G-invariant, forcing the denominator to be G-invariant as well. Thus L^G is precisely the field of fractions of R. This shows that R is a reasonably large subalgebra of S.

To summarize:

Proposition *Let V be a finite dimensional vector space over a field K of characteristic 0. Let G be a finite subgroup of $\mathrm{GL}(V)$, acting canonically on the symmetric algebra S of V^*, and let R be the subalgebra of G-invariants. Then the field of fractions of R coincides with the subfield of G-invariants in the field of fractions of S. In particular, it has transcendence degree n over K if V has dimension n.* □

3.2 Finite generation

Having seen in 3.1 that the ring of invariants R is not 'too small', we show next that it is finitely generated as a K-algebra (and therefore is not 'too big'). Our strategy (following Hilbert) is to exploit the fact that ideals in polynomial rings such as S are finitely generated (Hilbert's Basis Theorem); indeed, it follows easily from Hilbert's Theorem that a finite generating set can be extracted from any given set of generators. Accordingly, we consider the ideal $I = SR^+$ of S generated by the ideal R^+ of R consisting of elements with constant term 0. Choose a finite set of homogeneous generators for I from R^+. We shall show that any such set (together with 1) generates R as a K-algebra.

For the proof (and for later proofs), we need a sort of projection operator taking arbitrary elements of S into R, defined by 'averaging' over G. For any $f \in S$, define $f^\natural$ by the formula:

$$f^\natural := \frac{1}{|G|} \sum_{g \in G} g \cdot f. \tag{1}$$

It is clear that the assignment $f \mapsto f^\natural$ is a linear map of S onto R, preserving degrees and leaving all elements of R fixed. Of course, the fact that $|G|$ is not divisible by the characteristic of K is essential to the definition. Notice the following useful property of the averaging operator:

$$(pq)^\natural = p^\natural q \text{ whenever } p \in S, q \in R. \tag{2}$$

Proposition *With notation as above, suppose $f_1, \ldots, f_r$ are homogeneous elements of R^+ which generate the ideal $I = SR^+$ of S. Then R is generated as a K-algebra by these elements (together with 1).*

Proof. We have to show that every element $f \in R$ is a polynomial in $f_1, \ldots, f_r$. It is enough to do this for homogeneous elements f. Proceed by induction on the degree of f, the case of degree 0 being obvious. When $\deg f > 0$, we have $f \in I$, allowing us to write

$$f = s_1 f_1 + \ldots + s_r f_r, \text{ where } s_i \in S. \tag{3}$$

Since $f, f_1, \ldots, f_r$ are homogeneous, we may assume (after removing redundant terms from the s_i) that the s_i are also homogeneous, with $\deg s_i = \deg f - \deg f_i$ for all i. Next we apply our averaging operator to (3), recalling (2), to obtain

$$f = f^\natural = s_1^\natural f_1 + \ldots + s_r^\natural f_r.$$

Now the $s_i^\natural$ are homogeneous elements of R of degree less than $\deg f$, so by induction they are polynomials in $f_1, \ldots, f_r$, and then so is f. □

This proposition suggests a strategy for the proof of Chevalley's Theorem: select a minimal finite set of (homogeneous) generators for I from R^+, and go on to show that they must be algebraically independent.

Remark. Evidently the above proof of finite generation of R breaks down badly if we work over a field whose characteristic divides the order of G. The conclusion is still true, but requires a more subtle analysis due originally to E. Noether (see, e.g., Flatto [3]). In any case, our proof of finite generation is only qualitative; much more work is required to exhibit (for a given G) actual generators.

3.3   A divisibility criterion

Before launching into the proof of Chevalley's Theorem we record an easy lemma which will be needed a couple of times. Readers with some experience in algebraic geometry will see it as an immediate consequence of more sophisticated ideas (such as Hilbert's Nullstellensatz).

Lemma   *Let l be a homogeneous polynomial of degree 1 in the indeterminates $x_1, \ldots, x_n$, and suppose the polynomial f vanishes at all zeros of l. Then l divides f in the polynomial ring $K[x] = K[x_1, \ldots, x_n]$.*

Proof. Suppose (without loss of generality) that x_n occurs with a nonzero coefficient in l. Then we can carry out the usual division algorithm in one variable (without leaving $K[x]$) to obtain

$$f = lq + r, \tag{4}$$

where $q \in K[x]$ and r has degree 0 in x_n, i.e., $r \in K[x_1, \ldots, x_{n-1}]$. Unless $r = 0$ we get a contradiction as follows. Find elements $a_1, \ldots, a_{n-1}$ in the infinite field K for which $r(a_1, \ldots, a_{n-1}) \neq 0$. Plugging these values into l, we can solve a single linear equation to find a_n for which $l(a_1, \ldots, a_n) = 0$. By hypothesis, $f(a_1, \ldots, a_n) = 0$, contradicting (4). $\square$

3.4   The key lemma

For the rest of this chapter we return to the setting of Chapters 1 and 2: W is a finite (essential) group generated by reflections, acting on the n-dimensional euclidean space V over $\mathbf{R}$, which may be identified with $\mathbf{R}^n$. As in the preceding sections, W then acts naturally on the ring S of polynomial functions on V, which we identify with the ring $\mathbf{R}[x_1, \ldots, x_n]$. Again we denote by R the ring of W-invariants in S, and by I the ideal in S generated by R^+, the set of elements of R having constant term 0.

Exercise 1. Prove that W has an invariant of degree 2 in S. [Hint: W is a subgroup of the orthogonal group.]

We formulate as a lemma the step in the proof of Chevalley's Theorem which uses explicitly the fact that W is generated by reflections. The statement of this lemma obviously has something to do with dependence relations among polynomials, but is otherwise rather hard to motivate at this point. The reader will probably want to skip ahead to the next section, deferring the study of the lemma until it is actually needed.

Lemma *Let $f_1, \ldots, f_r \in R$, with f_1 not in the ideal of R generated by $f_2, \ldots, f_r$. Suppose $g_1, \ldots, g_r$ are homogeneous elements of S satisfying*

$$f_1 g_1 + \ldots + f_r g_r = 0. \tag{5}$$

Then $g_1 \in I$.

Proof. Observe first that f_1 cannot be in the ideal of S generated by $f_2, \ldots, f_r$. Otherwise we would have

$$f_1 = f_2 h_2 + \ldots + f_r h_r \text{ for some } h_i \in S. \tag{6}$$

Apply the averaging operator together with (2) of (3.2) to (6) to obtain

$$f_1 = f_1^\natural = f_2 h_2^\natural + \ldots + f_r h_r^\natural. \tag{7}$$

Since $h_i^\natural \in R$, equation (7) implies that f_1 is in the ideal of R generated by the other f_i, contradicting the hypothesis.

In order to prove that $g_1 \in I$, we use induction on $\deg g_1$. If g_1 is constant, it must be 0 (hence in I), since otherwise (5) would contradict the hypothesis on f_1. Now assume $\deg g_1 > 0$.

Consider a typical reflection $s = s_\alpha$ in W, and let l be a linear polynomial (uniquely determined up to a scalar multiple) whose zero set is the reflecting hyperplane H_α in $\mathbf{R}^n$. It is immediate that the polynomial $s \cdot g_i - g_i$ vanishes at all points of H_α, since $s = s^{-1}$ fixes such points. We can therefore invoke Lemma 3.3 to find polynomials h_i for which

$$s \cdot g_i - g_i = l h_i. \tag{8}$$

Both g_i and $s \cdot g_i$ are homogeneous (of the same degree), so (8) shows that h_i is also homogeneous and of lower degree than g_i. Now apply s to equation (5) to obtain:

$$f_1(s \cdot g_1) + \ldots + f_r(s \cdot g_r) = 0. \tag{9}$$

Subtract (5) from (9) and then substitute (8) to get

$$l(f_1 h_1 + \ldots + f_r h_r) = 0.$$

Since l is not identically zero, this in turn implies

$$f_1 h_1 + \ldots + f_r h_r = 0.$$

By induction, since $\deg h_1 < \deg g_1$, we get $h_1 \in I$. By (8), $s \cdot g_1 - g_1 \in I$, or $s \cdot g_1 \equiv g_1 (\bmod I)$.

Since W stabilizes R^+ and hence also I, it acts naturally on the quotient ring S/I. We have just seen that each reflection s acts trivially on the image of g_1; since the reflections generate W, this implies that $w \cdot g_1 \equiv g_1 (\bmod I)$ for all $w \in W$. In turn, $g_1^\sharp \equiv g_1 (\bmod I)$. But $g_1^\sharp$ is in R^+ (hence in I), forcing $g_1 \in I$ as desired. $\square$

Exercise 2. In the proof above, avoid the use of Lemma 3.3 to obtain equation (8) by working with an explicit expression for the action of s on V (hence on S).

3.5   Chevalley's Theorem

In this and later sections, it will be essential to work with partial derivatives of polynomials. As long as the base field is **R**, all the usual properties of partial differentiation for polynomial functions (notably the chain rule) may be invoked. Alternatively, one can develop these properties for polynomials in n indeterminates in a purely formal algebraic way (valid for any field). In any case, we shall need a familiar identity due to Euler for an arbitrary homogeneous polynomial $f(x_1, \ldots, x_n)$:

$$\sum_{i=1}^{n} x_i \frac{\partial f}{\partial x_i} = (\deg f)f. \tag{10}$$

The proof of this formula reduces at once to the special case when f is a monomial.

Theorem   *Let R be the subalgebra of $\mathbf{R}[x_1, \ldots, x_n]$ consisting of W-invariant polynomials. Then R is generated as an $\mathbf{R}$-algebra by n homogeneous, algebraically independent elements of positive degree (together with 1).*

Proof. As in 3.2 we consider the ideal I of S generated by the homogeneous invariants of positive degree. We can select (by Hilbert's Basis Theorem) a minimal generating set $f_1, \ldots, f_r$ for I consisting of homogeneous invariants of positive degree. Our main task is to show that these polynomials are algebraically independent. Once we do that, it will follow from Proposition 3.2 that (together with 1) they generate R as an algebra. In turn, it will follow from Proposition 3.1 that $r = n$, since the field of fractions of R must have transcendence degree n over **R**.

The proof of algebraic independence is a bit tricky. Suppose $f_1, \ldots, f_r$ are dependent, i.e., there exists a polynomial $h(y_1, \ldots, y_r) \neq 0$ for which

$$h(f_1, \ldots, f_r) = 0. \tag{11}$$

In order to keep track of degrees very precisely, we first refine the choice of h. Let

$$ay_1^{e_1} \cdots y_r^{e_r}$$

be any monomial occurring in h. If $d_i = \deg f_i$, set $d = \sum d_i e_i$, the degree of

$$af_1^{e_1} \cdots f_r^{e_r}$$

in $x_1, \ldots, x_n$. Evidently the various monomials in h which yield the same d add up to a nonzero polynomial with the same property (11) as h. So we may discard all other monomials in h.

Given the equation (11), it is reasonable to differentiate both sides with respect to x_k for each fixed k (using the chain rule):

$$\sum_{i=1}^{r} h_i \frac{\partial f_i}{\partial x_k} = 0, \quad \text{where } h_i = \frac{\partial h}{\partial y_i}(f_1, \ldots, f_r). \tag{12}$$

Note that h_i is a homogeneous element of R having degree $d - d_i$, while the $\partial f_i / \partial x_k$ are homogeneous elements of S. We would like to apply Lemma 3.4 to this situation, but unfortunately the hypothesis of that lemma might not be satisfied by the h_i. Renumber them if necessary so that $h_1, \ldots, h_m$ is a minimal generating set for the ideal of R generated by all of the h_i. Here $1 \le m \le r$. (If $m = r$, the rest of the argument will look much simpler.)

For each $i > m$, write

$$h_i = \sum_{j=1}^{m} g_{ij} h_j, \quad \text{where } g_{ij} \in R. \tag{13}$$

As a polynomial in $x_1, \ldots, x_n$, h_i is homogeneous of degree $d - d_i$, so after discarding redundant terms we can assume that each g_{ij} is homogeneous of degree $d_j - d_i$ $(= \deg h_i - \deg h_j)$. After substituting the equations (13) into (12), we obtain for each fixed k:

$$\sum_{i=1}^{m} h_i \left(\frac{\partial f_i}{\partial x_k} + \sum_{j=m+1}^{r} g_{ji} \frac{\partial f_j}{\partial x_k} \right) = 0. \tag{14}$$

Abbreviate the expression in parentheses by $p_i (1 \le i \le m)$ and note that p_i is homogeneous in $x_1, \ldots, x_n$ of degree $d_i - 1$.

Now we can apply Lemma 3.4 to (14) and conclude that $p_1 \in I$. Thus

$$\frac{\partial f_1}{\partial x_k} + \sum_{j=m+1}^{r} g_{j1} \frac{\partial f_j}{\partial x_k} = \sum_{i=1}^{r} f_i q_i, \tag{15}$$

where $q_i \in S$.

At first sight this does not appear helpful. But if we multiply both sides of (15) by x_k and sum over k, we can use Euler's formula (10) to get:

$$d_1 f_1 + \sum_{j=m+1}^{r} d_j g_{j1} f_j = \sum_{i=1}^{r} f_i r_i, \qquad (16)$$

where now $\deg r_i > 0$. The terms on the left side are homogeneous of degree d_1, so the term $f_1 r_1$ on the right side must cancel with other terms of degree different from d_1. After discarding all but terms of degree d_1, we see that (16) expresses f_1 as an element of the ideal in S generated by $f_2, \ldots, f_r$, contrary to the original choice of the f_i to be a minimal generating set of I. □

For brevity, we may refer to a set of algebraically independent homogeneous generators of R (of positive degree) as a set of **basic invariants** of R.

Exercise. State and prove a version of Chevalley's Theorem over an arbitrary field K of characteristic 0, defining a 'reflection' in $\mathrm{GL}(V)$ to be an element of order 2 which fixes a hyperplane pointwise.

Remark. When W is a Weyl group, a theorem of Harish–Chandra (see Humphreys [1], §23) allows one to derive from Chevalley's Theorem a description of the center of the universal enveloping algebra of a semisimple Lie algebra over $\mathbf{C}$: it too is a polynomial algebra.

3.6   The module of covariants

Our proof of Theorem 3.5 follows closely the original proof in Chevalley [2]. Implicit also in that paper is a complementary description of the R-module S, which follows readily from Lemma 3.4. While this result is inessential for what we do in the remainder of this chapter, it has an interesting cohomological interpretation (see the remark below). Moreover, it turns out to be equivalent to Chevalley's Theorem; for arrangements of the proof emphasizing this equivalence, see Bourbaki [1], V, §5, Hiller [3], II.3, Springer [3], 4.2.

Proposition   *Viewed as an R-module, S is free of rank $|W|$.*

Proof. The idea is to compare the R-module S with the vector space S/I, where as before I denotes the ideal generated by the homogeneous invariants of positive degree. Start with a set of homogeneous polynomials $g_\alpha \in S$ whose cosets $g_\alpha + I$ span S/I. We claim the g_α must span the R-module S. Clearly the submodule T which they do span is graded by degree, so it is enough to show by induction on d that $T_d = S_d$. Since $S_0 \cap I = 0$, some g_α must have degree 0, so $T_0 = S_0$. Next take

$f \in S_d, d > 0$, and write it as a finite linear combination

$$f = \sum_\alpha c_\alpha g_\alpha + \sum_\beta f_\beta h_\beta,$$

where $c_\alpha \in \mathbf{R}$, $h_\beta \in R$ has positive degree, and f_β is homogeneous of degree less than d. By induction, all $f_\beta \in T$, forcing $f \in T$.

Now suppose $g_1, \ldots, g_m$ are homogeneous elements of S whose cosets $g_i + I$ are linearly independent in S/I. Using induction on m, we show that these elements are independent in the R-module S, the case $m = 1$ being clear. If there is an R-linear combination of the g_i equalling 0, there must be such a relation

$$f_1 g_1 + \ldots + f_m g_m = 0$$

with all f_i homogeneous in R. Since $g_1 \notin I$, we can appeal to Lemma 3.4 to conclude that

$$f_1 = h_2 f_2 + \ldots + h_m f_m$$

for some (homogeneous) elements $h_i \in R$. After substituting, this yields:

$$f_2(g_2 + h_2 g_1) + \ldots + f_m(g_m + h_m g_1) = 0.$$

Note that the $g_i + h_i g_1$ are homogeneous, and their cosets are linearly independent in S/I. By induction, $f_2, \ldots, f_m$ are all 0, and in turn $f_1 = 0$.

Combining these steps, we see that a vector space basis of S/I leads to an R-module basis of S. Recalling the well-known exercise below, this in turn immediately yields a basis for the extension of the field of fractions of S over the field of fractions of R, which we know has dimension $|W|$ (3.1). $\square$

Exercise. Let B be a subring (with 1) of the integral domain A; denote the respective fields of fractions by F and E. Suppose E/F is a finite extension. If the B-module A is free of rank r, then $[E : F] = r$. (In fact, a module basis of A over B is also a vector space basis of E over F.)

The proof shows that the vector space S/I has dimension $|W|$. Moreover, W acts naturally as a group of linear operators on this space. It can be shown without too much difficulty that this representation of W is equivalent to the *regular representation*, using some standard information from Galois theory (Normal Basis Theorem) to analyze the action of W on the field of fractions of S. (See Chevalley [2] or Bourbaki [1], V, 5.2.) The action of W preserves the natural grading of S/I, and the decomposition of the graded pieces turns out to be very interesting (see Beynon–Lusztig [1]).

Remark. The study of the 'coinvariant algebra' S/I is a central theme of Hiller [3]. When W is a Weyl group, its invariant theory is closely related to the topology of the corresponding compact Lie group and its 'flag manifold'. Early work of A. Borel showed how to identify the cohomology algebra of the flag manifold with S/I; this also has an interpretation in terms of the Bruhat cell decomposition (indexed by W). See Bernstein–Gelfand–Gelfand [1], Demazure [1], Hiller [1]–[3].

3.7　Uniqueness of the degrees

The algebraically independent generators of R provided by Theorem 3.5 need not be uniquely determined, e.g., $x_1 + x_2$ and $x_1^2 + x_2^2$ work just as well as the elementary symmetric polynomials $x_1 + x_2$ and $x_1 x_2$. However, the *degrees* do turn out to be independent of the choice of generators.

Proposition　*Suppose that $f_1, \ldots, f_n$ and $g_1, \ldots, g_n$ are two sets of homogeneous, algebraically independent generators of the ring R of W-invariant polynomials. Denote the respective degrees by d_i and e_i. Then, after renumbering one of the sets if necessary, we have $d_i = e_i$ for all i.*

Proof. Each set of polynomials can be written as polynomials in the other set. For each pair of indices (i, j), we can use the chain rule to evaluate the partial derivative $\partial f_i / \partial f_j$:

$$\sum_{k=1}^{n} \frac{\partial f_i}{\partial g_k} \frac{\partial g_k}{\partial f_j} = \delta_{ij}.$$

This shows that the matrices

$$\left(\frac{\partial f_i}{\partial g_j} \right) \text{ and } \left(\frac{\partial g_i}{\partial f_j} \right)$$

are inverses, and therefore each has nonzero determinant. The expansion of the first determinant as a sum of signed products must involve a nonzero product

$$\prod_{i=1}^{n} \frac{\partial f_i}{\partial g_{\pi(i)}}$$

for some permutation π. After renumbering the g_i we may assume that π is the identity. Thus when f_i is written as a polynomial in $g_1, \ldots, g_n$, g_i must actually occur. After discarding redundant terms, we may assume that each monomial

$$g_1^{k_1} \cdots g_n^{k_n}$$

occurring in f_i satisfies: $d_i = \sum e_j k_j$. So $d_i \geq e_i$. In turn,

$$\sum_{i=1}^n d_i \geq \sum_{i=1}^n e_i.$$

By interchanging the role of the f_i and g_i, the same argument produces the reverse inequality. Finally, we conclude that $d_i = e_i$ for all i. $\square$

For brevity, we may refer to the numbers $d_1, \ldots, d_n$ (usually written in increasing order of magnitude) as the **degrees** of W. The rest of this chapter will be largely devoted to studying their remarkable properties and to computing them in all cases. It was noted as an exercise in 3.4 that W must have an invariant of degree 2: as a group of orthogonal transformations, it leaves invariant the polynomial $x_1^2 + \ldots + x_n^2$ when V is identified with $\mathbf{R}^n$ with its usual euclidean structure. (Of course, the degree 1 would also occur if W failed to be essential.)

Exercise. The scalar transformation -1 lies in W if and only if all degrees are even. [-1 induces an automorphism of S, acting on S_d as $(-1)^d$. One implication is easy, but the other may require some Galois theory.]

Type	$d_1, \ldots, d_n$
A_n	$2, 3, \ldots, n+1$
B_n	$2, 4, 6, \ldots, 2n$
D_n	$2, 4, 6, \ldots, 2n-2, n$
E_6	$2, 5, 6, 8, 9, 12$
E_7	$2, 6, 8, 10, 12, 14, 18$
E_8	$2, 8, 12, 14, 18, 20, 24, 30$
F_4	$2, 6, 8, 12$
G_2	$2, 6$
H_3	$2, 6, 10$
H_4	$2, 12, 20, 30$
$I_2(m)$	$2, m$

Table 1: Degrees of basic invariants

After some further generalities, we shall be better able to discuss concrete examples (3.12). At the risk of lessening the suspense, we provide in Table 1 the list of degrees for each type of irreducible W. It will be some time before we succeed in verifying the table completely, but meanwhile the reader can compare it with the partial results obtained. (Notice that, when n is even, the degree n occurs twice in the list for D_n. This is the only case involving such a repetition.)

3.8   Eigenvalues

The study of invariants may be viewed as the study of eigenspaces of certain operators for the eigenvalue 1. But a deeper probe of the degrees of W requires some consideration of all eigenvalues, together with related traces and determinants. First we note a convenient description of the dimension of the space of W-invariants in an arbitrary linear representation. (Here W could be any finite group.)

Lemma   *Let E be any finite-dimensional W-module over a field of characteristic 0. Then the dimension of the space of W-invariants in E is given by the trace of the linear operator on E defined by*

$$z = \frac{1}{|W|} \sum_{w \in W} w. \tag{17}$$

Proof. Note that $wz = z$ for all $w \in W$. Using this, a quick calculation shows that the operator z is idempotent. Thus it is diagonalizable (with possible eigenvalues $0,1$), since its minimal polynomial divides $x^2 - x$ and therefore has distinct roots. Let $E = E_0 \oplus E_1$ be the eigenspace decomposition (possibly one of these subspaces is 0). It is clear that the trace of z is $\dim E_1$. But E_1 is the space of all W-invariants in E. Say $e \in E_1$. Then $e = z \cdot e = wz \cdot e = w \cdot e$ for all $w \in W$. In the other direction, if e is W-invariant, then

$$z \cdot e = \frac{1}{|W|} \sum w \cdot e = \frac{1}{|W|} \sum e = e,$$

so $e \in E_1$. □

Now we can develop a combinatorial identity of the frequently occurring type 'sum = product', involving the degrees of W on the 'product' side. The 'sum' is a formal power series, involving the action of W on S.

In order to work explicitly with eigenvalues, we have to extend the base field from $\mathbf{R}$ to $\mathbf{C}$. As an element of finite order in $\mathrm{GL}(V)$, each $w \in W$ acts via a diagonal matrix relative to a suitable basis of $V_{\mathbf{C}}$. Moreover, the eigenvalues of w are roots of unity; they are real or else occur in complex conjugate pairs (since w is represented by a real matrix). The eigenvalues of w on V^* are just the reciprocals ($=$ complex conjugates) and therefore are the same as the eigenvalues on V. Now if t is a complex number, it makes sense to write

$$\det(1 - tw) = (1 - c_1 t) \cdots (1 - c_n t), \tag{18}$$

where w has eigenvalues $c_1, \ldots, c_n$. We can also regard t as an indeterminate (by formally extending the base field to $\mathbf{C}(\mathrm{t})$), in which case the reciprocal of (18) has an expansion as a formal power series:

$$\frac{1}{\det(1-tw)} = (1 + c_1 t + c_1^2 t^2 + \dots) \cdots (1 + c_n t + c_n^2 t^2 + \dots)$$

$$= \sum_{k \geq 0} \left(\sum_{k_1 + \dots + k_n = k} c_1^{k_1} \cdots c_n^{k_n} \right) t^k. \tag{19}$$

Proposition *Viewing both sides as formal power series in t, we have:*

$$\frac{1}{|W|} \sum_{w \in W} \frac{1}{\det(1-tw)} = \prod_{i=1}^{n} \frac{1}{(1-t^{d_i})}. \tag{20}$$

Proof. Fix $w \in W$, with eigenvalues c_i as above. Earlier we viewed S as the algebra of polynomials in $x_1, \dots, x_n$ (a basis of V^*). After extending the base field to $\mathbf{C}$, we can instead work with a basis $z_1, \dots, z_n$ of the complexified dual space consisting of eigenvectors for w. To compute the eigenvalues of w on the homogeneous component S_k of S, we can use the basis of the complexified space consisting of monomials

$$z_1^{k_1} \cdots z_n^{k_n}, \quad \text{where } k_1 + \dots + k_n = k.$$

These are eigenvectors for w corresponding to the eigenvalues $c_1^{k_1} \cdots c_n^{k_n}$. The sum of these eigenvalues is the trace of w on S_k, and agrees with the coefficient of t^k in the power series (19).

In view of this interpretation of (19), the coefficient of t^k in the left hand side of (20) is the trace of the linear operator

$$\frac{1}{|W|} \sum_{w \in W} w$$

on S_k. By the above lemma, this is precisely the dimension of the space R_k of homogeneous invariants of degree k. But the dimension of this space can be computed another way. If $f_1, \dots, f_n$ is a basic set of invariants, of degrees $d_1, \dots, d_n$, the monomials

$$f_1^{e_1} \cdots f_n^{e_n} \quad \text{with } \sum d_i e_i = k$$

form a basis of R_k. The number of such n-tuples $(e_1, \dots, e_n)$ is evidently the coefficient of t^k in the formal power series

$$(1 + t^{d_1} + t^{2d_1} + \dots) \cdots (1 + t^{d_n} + t^{2d_n} + \dots),$$

which is the same as the product on the right hand side of (20). □

3.9   Sum and product of the degrees

From the identity in Proposition 3.8 we can easily derive computable expressions for the sum and the product of the degrees of W. Since the trace of each $w \in W$ is real, it is clear that the only elements of W having $n-1$ eigenvalues equal to 1 are the identity and the N reflections (where N is the number of positive roots, by 1.14). Thus the polynomial $\det(1 - tw)$ is equal to $(1-t)^n$ if $w = 1$, or is equal to $(1-t)^{n-1}(1+t)$ if w is a reflection, but is otherwise not divisible by $(1-t)^{n-1}$.

Theorem   *Let $d_1, \ldots, d_n$ be the degrees of W, and N the number of reflections in W. Then $d_1 d_2 \cdots d_n = |W|$ and $d_1 + d_2 + \ldots + d_n = N + n$.*

Proof. Multiply both sides of (20) in Proposition 3.8 by $(1-t)^n$ to obtain

$$\frac{1}{|W|} \left(1 + N \frac{(1-t)}{(1+t)} + (1-t)^2 g(t) \right) = \prod_{i=1}^{n} \frac{1}{1 + t + \ldots + t^{d_i - 1}} \qquad (21)$$

Here $g(t)$ is a rational function with denominator not divisible by $1 - t$. Set $t = 1$ to get

$$\frac{1}{|W|} = \frac{1}{d_1 \cdots d_n},$$

or $|W| = d_1 \cdots d_n$.

If instead we (formally) differentiate both sides of (21), we get

$$-\frac{2N}{|W|} \frac{1}{(1+t)^2} + h(t) =$$

$$\left(\prod_{i=1}^{n} \frac{1}{1 + t + \ldots + t^{d_i - 1}} \right) \left(\sum_{i=1}^{n} -\frac{1 + 2t + \ldots + (d_i - 1)t^{d_i - 2}}{1 + t + \ldots + t^{d_i - 1}} \right), \qquad (22)$$

where $h(t)$ is a rational function with numerator divisible by $1 - t$. Now set $t = 1$ in (22) to obtain

$$-\frac{N}{2|W|} = -\frac{1}{2} \frac{1}{d_1 \cdots d_n} \sum_{i=1}^{n} (d_i - 1).$$

Substituting for $|W|$ the product of the degrees, this yields the desired expression for the sum of the degrees. $\square$

When W is a symmetric group, the reader should have no trouble verifying these formulas directly. When $W = \mathcal{D}_m$, the degrees d_1, d_2 satisfy $d_1 d_2 = 2m$ and $d_1 + d_2 = m + 2$, forcing $d_1 = 2, d_2 = m$.

Exercise. Check that the degrees listed in Table 1 of 3.7 are compatible with the theorem above, referring to Table 2 in 2.11.

In order to treat more complicated examples, we need to develop (in the following section) an effective way to test a given set of polynomials for algebraic independence. This will also make it easy to prove a theorem of Shephard and Todd (3.11) which asserts that only for reflection groups can the ring of invariants be generated by algebraically independent polynomials.

For use in the proof of Theorem 3.11, we make a simple but important observation. In 3.8 as well as in the proof of the above theorem, the only fact about W used in an essential way is that its ring of invariants is generated by algebraically independent homogeneous polynomials. (Check this!)

Remark. It is natural to wonder whether the other elementary symmetric polynomials in $d_1, \ldots, d_n$ can be interpreted in an interesting way. Actually, it is better to consider the related numbers $d_i - 1$ (whose significance will become clear later in the chapter). The theorem shows that their sum is N, the number of reflections in W. It was observed by Shephard–Todd [1] that

$$\prod_{i=1}^{n}(1 + (d_i - 1)t) = a_0 + a_1 t + \cdots + a_n t^n,$$

where a_k is the number of elements of W whose fixed point space in V has dimension $n - k$. Note that setting $t = 1$ in the formula recovers the fact that $|W|$ is the product of the d_i. A uniform proof of the formula, involving study of the ring of invariant differential forms, was later given by Solomon [1] (see the review by Steinberg, *Math. Reviews* **27** #4872). Following work of V.I. Arnol'd on symmetric groups, Brieskorn [1] gave a nice topological interpretation of the formula: the left side is the Poincaré polynomial of the complement of reflecting hyperplanes (in the complexified setting), cf. Lehrer [2], Orlik–Solomon [2].

3.10 Jacobian criterion for algebraic independence

There is a simple criterion for the algebraic independence (over an arbitrary field of characteristic 0) of n polynomials $f_1, \ldots, f_n$ in n indeterminates $x_1, \ldots, x_n$, expressed in terms of the Jacobian determinant. Write $J(f_1, \ldots, f_n)$ for the determinant of the $n \times n$ matrix whose (i, j)-entry is $\partial f_i / \partial x_j$.

Proposition *The polynomials $f_1, \ldots, f_n$ in indeterminates $x_1, \ldots, x_n$ are algebraically independent (over a field K of characteristic 0) if and only if $J(f_1, \ldots, f_n) \neq 0$.*

Proof. One implication is straightforward. Suppose the polynomials are algebraically dependent, so that $h(f_1, \ldots, f_n) = 0$ for some nonzero polynomial $h(y_1, \ldots, y_n)$. We may assume that the degree of h is as small as possible. For each fixed j, differentiate this relation with respect to x_j (using the chain rule) to get an equation:

$$\sum_{i=1}^{n} \frac{\partial h}{\partial y_i}(f_1, \ldots, f_n) \frac{\partial f_i}{\partial x_j} = 0. \qquad (23)$$

The equations (23) for $1 \leq j \leq n$ form a system of linear equations over the field $K(x_1, \ldots, x_n)$ with coefficient matrix of determinant $J(f_1, \ldots, f_n)$ and with 'unknowns'

$$\frac{\partial h}{\partial y_i}(f_1, \ldots, f_n). \qquad (24)$$

Because h is not constant, not all of the partial derivatives $\partial h/\partial y_i$ can vanish; since each has smaller degree than h, the choice of h shows that the polynomials (24) cannot all be 0. Thus the linear system has a nontrivial solution, forcing its coefficient matrix to have determinant 0.

The reverse implication is less transparent. Suppose $f_1, \ldots, f_n$ are algebraically independent. Since $K(x_1, \ldots, x_n)$ has transcendence degree n over K, the polynomials $x_i, f_1, \ldots, f_n$ are algebraically dependent for each fixed i. Let $h_i(y_0, y_1, \ldots, y_n)$ be a polynomial of minimal positive degree for which

$$h_i(x_i, f_1, \ldots, f_n) = 0. \qquad (25)$$

Now differentiate (25) with respect to x_k to obtain:

$$\sum_{j=1}^{n} \frac{\partial h_i}{\partial y_j}(x_i, f_1, \ldots, f_n) \frac{\partial f_j}{\partial x_k} + \frac{\partial h_i}{\partial y_0}(x_i, f_1, \ldots, f_n)\delta_{ik} = 0. \qquad (26)$$

Since the f_j are algebraically independent, h_i must have positive degree in y_0. So $\partial h_i/\partial x_i$ is nonzero and of smaller degree than h_i, forcing the value of this polynomial at $x_i, f_1, \ldots, f_n$ to be nonzero. Transpose these terms to the right side of the equations (26) for $1 \leq i, k \leq n$, and write the equations in matrix form as

$$\left(\frac{\partial h_i}{\partial y_j}\right)\left(\frac{\partial f_i}{\partial x_j}\right) = \left(-\delta_{ij}\frac{\partial h_i}{\partial x_j}\right). \qquad (27)$$

The matrix on the right side of (27) is a diagonal matrix with nonzero determinant, so the Jacobian determinant on the left side is also nonzero. $\square$

Corollary   *Suppose $f_1, \ldots, f_n$ are algebraically independent and homogeneous, of respective degrees $d_1, \ldots, d_n$. Then $J(f_1, \ldots, f_n)$ is homogeneous of degree $\sum(d_i - 1) = N$.*

Proof. The above proposition shows that the Jacobian is nonzero. It can be expressed as a sum of signed products, each product being of the form

$$\frac{\partial f_1}{\partial x_{\pi(1)}} \cdots \frac{\partial f_n}{\partial x_{\pi(n)}}$$

for some permutation π of the indices. For each nonzero product of this type, the individual terms are nonzero and homogeneous, of respective degrees $d_1-1, \ldots, d_n-1$. In turn, thanks to Theorem 3.9, $\sum(d_i-1) = N$. □

Exercise. Set $f_k(x) = \sum_i x_i^k$ for each $k = 1, \ldots, n$. Verify that $f_1, \ldots, f_n$ are algebraically independent.

3.11 Groups with free rings of invariants

There is a sort of converse to Theorem 3.5, proved by Shephard–Todd [1]. While it is not essential to the further study of degrees in this chapter, it helps to underscore the special status of finite reflection groups among all finite linear groups.

Theorem *Let V be an n-dimensional euclidean space over $\mathbf{R}$, and let G be a finite subgroup of $\mathrm{GL}(V)$, acting naturally on the polynomials $S = \mathbf{R}[x_1, \ldots, x_n]$. Suppose the ring of invariants S^G is generated by n algebraically independent homogeneous polynomials $g_1, \ldots, g_n$. Then G is generated by the reflections it contains.*

Proof. Denote by H the (possibly trivial) subgroup of G generated by the reflections in G. Theorem 3.5 says that the ring S^H is generated by n algebraically independent homogeneous polynomials $f_1, \ldots, f_n$. Say $\deg f_i = d_i$ and $\deg g_i = e_i$. Evidently $S^G \subset S^H$, so the g_i can be written as polynomials in the f_i. After discarding redundant terms, we may assume that each monomial

$$f_1^{k_1} \cdots f_n^{k_n}$$

occurring in g_i satisfies: $e_i = \sum d_j k_j$.

Now we use an argument similar to that in the proof of Proposition 3.7 to compare degrees. Use the chain rule to differentiate:

$$\frac{\partial g_i}{\partial x_k} = \sum_j \frac{\partial g_i}{\partial f_j} \frac{\partial f_j}{\partial x_k}. \tag{28}$$

Thanks to Proposition 3.10, the Jacobian determinant involving the $\partial g_i / \partial x_k$ is nonzero, so the corresponding Jacobian involving the $\partial g_i/\partial f_j$ on the right side of (28) must be nonzero. After renumbering if necessary, we can assume that a product of the form

$$\frac{\partial g_1}{\partial f_1} \cdots \frac{\partial g_n}{\partial f_n}$$

is nonzero. In turn, this forces $e_i \geq d_i$ for all i. As observed in 3.9, Theorem 3.9 can be applied to G as well as H. Therefore

$$\sum(d_i - 1) = N = \sum(e_i - 1), \tag{29}$$

where N is the number of reflections in H (= the number of reflections in G). Thus $e_i = d_i$ for all i. But Theorem 3.9 also shows that

$$|G| = \prod e_i \text{ and } |H| = \prod d_i,$$

forcing $G = H$. $\square$

Remark. This theorem, as well as Chevalley's Theorem, is actually valid in a wider setting. Define an endomorphism of a finite-dimensional vector space over $\mathbf{C}$ to be a **pseudo-reflection** if it has finite order and its fixed point space is of codimension 1; see Shephard [1]. (This is sometimes called a 'unitary' or 'complex' reflection, or just a 'reflection'.) The previous arguments apply with only minor changes to the ring of polynomial invariants of a finite group generated by pseudo-reflections (called a **unitary reflection group** or **complex reflection group**). In particular, one has well-determined degrees whose product is the group order. Among these groups are the reflection groups we have been studying (complexified). But in general the complex reflection groups are not Coxeter groups in any obvious way, so we do not pursue them here. They were classified using geometric methods by Shephard–Todd [1]. Suggestions of Coxeter [5] led Cohen [1] to a more elegant formulation, using 'root graphs'. (Cohen [2] has also studied quaternionic reflection groups.)

The formula of Shephard–Todd in the remark at the end of 3.9 remains valid in the setting of groups generated by pseudo-reflections. There has also been much interest in the topology of the complement of hyperplanes for a complex reflection group. See the papers of Orlik–Solomon as well as Terao [1] for further results.

3.12 Examples

We return to the study of a finite (real) reflection group W. In some cases it is easy to exhibit a set of basic invariants and thereby compute the degrees directly. This approach relies on the following criterion.

Proposition *Suppose $g_1, \ldots, g_n$ are homogeneous W-invariants, having respective degrees $e_1, \ldots, e_n$. If $g_1, \ldots, g_n$ are algebraically independent and $\prod e_i = |W|$, then they are a set of basic invariants.*

Proof. We may assume that $e_1 \leq e_2 \leq \ldots \leq e_n$. Let $f_1, \ldots, f_n$ be a set of basic invariants, of degrees $d_1 \leq d_2 \leq \ldots \leq d_n$. Since g_1 is a polynomial in the f_i, it is clear that $e_1 \geq d_1$. We claim that this inequality holds for each i. Otherwise, let k be the first index for which $e_k < d_k$. Then each of $g_1, \ldots, g_k$ must be a polynomial in $f_1, \ldots, f_{k-1}$. But the field of rational functions generated by $g_1, \ldots, g_k$ has transcendence degree k over $\mathbf{R}$, so cannot be contained in a field of smaller transcendence degree. This proves our claim.

Thanks to Theorem 3.9 and the hypothesis, $\prod d_i = |W| = \prod e_i$, forcing $d_i = e_i$ for all i. In turn, we see that the dimension of the space of homogeneous invariants of degree d generated by the g_i agrees with that of the space generated by the f_i, for every d. Thus the g_i are a set of basic invariants for W. $\square$

We apply this criterion to groups of types A_n, B_n, D_n. From the description of Φ in 2.10, one sees immediately how W acts on the polynomial functions.

Consider first the symmetric group $W = \mathcal{S}_{n+1}$, of type A_n. Since W is required to be essential, it acts by permuting $x_1, \ldots, x_{n+1}$ subject to the relation $x_{n+1} = -(x_1 + \ldots + x_n)$. Rather than work with elementary symmetric polynomials, we let

$$f_i := x_1^{i+1} + \ldots + x_{n+1}^{i+1} \quad (1 \leq i \leq n).$$

(These are related to the elementary symmetric polynomials by Newton's identities.) The product of degrees of the f_i is $(n+1)! = |W|$, so it just has to be checked (using the Jacobian criterion) that the f_i are algebraically independent. For $1 \leq i, j \leq n$,

$$\frac{\partial f_i}{\partial x_j} = (i+1)x_j^i - (i+1)x_{n+1}^i.$$

Thus $J = J(f_1, \ldots, f_n)$ is $(n+1)!$ times the $n \times n$ determinant K with (i, j) entry $x_j^i - x_{n+1}^i$, which in turn is a close relative of the familiar Vandermonde determinant $V =$

$$\begin{vmatrix} 1 & \cdots & 1 \\ x_1 & \cdots & x_{n+1} \\ \cdot & & \cdot \\ \cdot & & \cdot \\ \cdot & & \cdot \\ x_1^n & \cdots & x_{n+1}^n \end{vmatrix}$$

Now V is well-known to equal

$$\prod_{1 \leq i < j \leq n+1} (x_j - x_i).$$

By subtracting the last column of V from each of the earlier columns, one gets $V = (-1)^n K$. Expressed in terms of $x_1, \ldots x_n$, we see that

$$J = (n+1)! \prod_{1 \leq i < j \leq n} (x_j - x_i) \prod_{i=1}^{n} (x_i + z),$$

where $z := x_1 + \ldots + x_n$. Thus $J \neq 0$.

For W of type B_n the reasoning is similar. Here W acts on $x_1, \ldots, x_n$ by permutations and sign changes, leaving invariant

$$f_i := x_1^{2i} + \ldots + x_n^{2i} \quad (1 \leq i \leq n),$$

whose degrees have product $2^n n! = |W|$. A quick computation yields

$$J = 2^n n! \, x_1 \cdots x_n \prod_{1 \leq i < j \leq n} (x_j^2 - x_i^2) \neq 0.$$

(Note that we might also have chosen as basic invariants the elementary symmetric polynomials in the squares of the variables.)

The group W of type D_n acts on $x_1, \ldots, x_n$ by permutations and by changes of an even number of signs, so we can easily find invariants by modifying the preceding choice slightly:

$$f_i := \sum_{j=1}^{n} x_j^{2i} \quad (1 \leq i \leq n-1), \quad f_n := x_1 \cdots x_n.$$

The product of the degrees is $2^{n-1} n! = |W|$. With somewhat more effort than before, one finds

$$J = (-2)^{n-1}(n-1)! \prod_{1 \leq i < j \leq n} (x_j^2 - x_i^2).$$

Exercise. Find basic invariants for the dihedral groups.

3.13 Factorization of the Jacobian

In preparation for a deeper study of the degrees of W, we have to look more closely at the Jacobian determinant

$$J = J(f_1, \ldots, f_n) = \det\left(\frac{\partial f_i}{\partial x_j}\right),$$

where $f_1, \ldots, f_n$ is a set of basic invariants. So far we have only been concerned with the fact that the Jacobian is not identically zero. We can actually describe how J factors in S and what its precise role there is. To this end, let l_α ($\alpha \in \Phi$) be a linear polynomial whose zero set is the hyperplane H_α in V orthogonal to α. Since l_α is only determined up to a nonzero scalar multiple, we specify it (in coordinate-free fashion) to be the map: $\lambda \mapsto (\alpha, \lambda)$. A quick calculation shows that for all roots α, β: $s_\beta \cdot l_\alpha = l_{s_\beta \alpha}$. Also, $l_{-\alpha} = -l_\alpha$.

Define a polynomial $f \in S$ to be **alternating** if $w \cdot f = (\det w)f$ for all $w \in W$. (The terms 'anti-invariant' and 'skew-invariant' may also be met in the literature.) The alternating polynomials form a subspace A of S, which is clearly the direct sum of its homogeneous components $A_k := A \cap S_k$.

Proposition *Fix a set of basic invariants $f_1, \ldots, f_n$ for W, and let J be the corresponding Jacobian. For each root $\alpha \in \Phi$, define l_α as above, so that its zero set is the orthogonal hyperplane H_α.*

(a) *$J = k \prod_{\alpha \in \Phi^+} l_\alpha$ for some $k \in \mathbf{R}$ (depending on the choice of the f_i).*

(b) *A polynomial $f \in S$ is alternating if and only if it can be written as the product of J and an invariant polynomial.*

(c) *For each k, $\dim A_k = \dim R_{k-N}$.*

Proof. (a) Define a mapping $\varphi : \mathbf{R}^n \to \mathbf{R}^n$ by setting

$$\varphi(a_1, \ldots, a_n) = (f_1(a_1, \ldots, a_n), \ldots, f_n(a_1, \ldots, a_n)).$$

Suppose $a = (a_1, \ldots, a_n) \in H_\alpha$ for some root α. Then every open neighborhood of a contains a pair of distinct points b, c for which $s_\alpha b = c$. But then $f_i(c) = f_i(s_\alpha b) = (s_\alpha \cdot f_i)(b) = f_i(b)$, forcing $\varphi(c) = \varphi(b)$. According to the Inverse Function Theorem, for every point a at which J does not vanish, φ maps some open neighborhood of a one-to-one onto some open neighborhood of $\varphi(a)$. We conclude that J must vanish on H_α for all $\alpha \in \Phi$. Thanks to Lemma 3.3, l_α divides J. Since the irreducible polynomials $l_\alpha, \alpha \in \Phi^+$ are nonproportional, their product also divides J. But this product has degree N, which is also the degree of J according to Theorem 3.9 and Corollary 3.10. (For a purely algebraic proof, avoiding the Inverse Function Theorem, see Flatto [3], p. 253.)

(b) In view of (a), we can assume without loss of generality that

$$J = \prod_{\alpha \in \Phi^+} l_\alpha.$$

It has to be shown that a polynomial is alternating if and only if it is the product of an invariant polynomial with J. We check first that J really is alternating, using the fact that $s_\beta \cdot l_\alpha = l_{s_\beta \alpha}$. When β is simple,

s_β maps β to its negative and permutes the other positive roots (1.4). Therefore

$$s_\beta \cdot \prod_{\alpha \in \Phi^+} l_\alpha = - \prod_{\alpha \in \Phi^+} l_\alpha.$$

Iterating this calculation, we get $w \cdot J = (\det w)J$ as desired. Moreover, it is obvious that the product of J with any invariant is alternating.

Now take f to be any alternating polynomial, so $s_\alpha \cdot f = -f$ for any reflection s_α. If $a \in H_\alpha$ it follows that

$$-f(a) = (s_\alpha \cdot f)(a) = f(s_\alpha(a)) = f(a),$$

forcing $f(a) = 0$. Since f vanishes on the zero set of l_α, Lemma 3.3 says that l_α divides f. As in the proof of (a) above, it follows that J divides f: $f = gJ$ for some $g \in S$. Applying an arbitrary $w \in W$ to both sides, we get: $(\det w)f = w \cdot f = (w \cdot g)(w \cdot J) = (\det w)(w \cdot g)J$, and therefore $w \cdot g = g$. So f is the product of J with an element of R.

(c) This follows immediately from (b), since J has degree N. $\square$

We shall use part (c) right away in 3.15, while part (a) is needed again in 3.19.

Example. Consider again the group of type $\mathbf{B}_n$ discussed in 3.12. The factorization of J found there agrees with the above proposition, since each of the linear functions $x_i = 0$ defines a hyperplane orthogonal to one of the short roots $\pm \varepsilon_i$ and each of the factors $x_j + x_i$ or $x_j - x_i$ defines a hyperplane orthogonal to a long root $\pm(\varepsilon_j - \varepsilon_i)$ or $\pm(\varepsilon_j + \varepsilon_i)$; see 2.10.

3.14 Induction and restriction of class functions

In preparation for the theorem in the following section, we have to recall some simple facts about **class functions** on finite groups (that is, **C**-valued functions which are constant on conjugacy classes). Fix a finite group G and subgroup H. If $\chi : G \to \mathbf{C}$ is a class function on G, it is obvious that the restriction to H (denoted χ_H) is a class function on H. In the other direction, given a class function φ on H, we obtain an **induced** class function φ^G by setting

$$\varphi^G(g) := \frac{1}{|H|} \sum_{x \in G} \varphi(xgx^{-1}),$$

where the sum runs over those $x \in G$ for which $xgx^{-1} \in H$.

In the special case $\varphi = 1_H$ (taking value 1 at all elements of H), the induced class function has a useful interpretation: $1_H^G(g)$ is $1/|H|$ times the number of $x \in G$ with $xgx^{-1} \in H$, or $gx^{-1}H = x^{-1}H$. So $1_H^G(g)$ is the number of distinct left cosets $x^{-1}H$ fixed by g. Recalling the set-up of 1.16, we can therefore describe the function f_I on W introduced there as the class function induced from 1_{W_I}. Using this language, we can reformulate Proposition 1.16 as follows:

$$\sum_{I \subset S} (-1)^I 1_{W_I}^W(w) = \det(w) \quad \text{for all } w \in W.$$

In the next section we shall need two further observations about induction and restriction:

Lemma   (a) *If χ is any class function on G, then $\chi(1_H^G) = (\chi_H)^G$.*
(b) *If φ is any class function on H, then*

$$\frac{1}{|G|} \sum_{g \in G} \varphi^G(g) = \frac{1}{|H|} \sum_{h \in H} \varphi(h).$$

Proof. (a) We calculate as follows:

$$
\begin{aligned}
(\chi_H)^G(g) &= \frac{1}{|H|} \sum \chi_H(xgx^{-1}) \\
&= \frac{1}{|H|} \sum \chi(g) \\
&= \chi(g) \frac{1}{|H|} \sum 1 \\
&= \chi(g)\, 1_H^G(g).
\end{aligned}
$$

In each step the summation is taken over those $x \in G$ for which $xgx^{-1} \in H$. In the second equality we used the fact that χ is a class function on G, and in the final equality we used the definition of induction.

(b) By definition, the left side involves a double summation over those $(g, x) \in G \times G$ for which $xgx^{-1} \in H$, divided by $|G||H|$. A little bookkeeping shows that each element of G (in particular, each element of H) occurs in the form xgx^{-1} for $|G|$ distinct pairs (g, x). This yields the right hand side. $\square$

3.15   Factorization of the Poincaré polynomial

By combining a number of the previous results, we can now obtain a beautiful factorization of the Poincaré polynomial $W(t) = \sum_{w \in W} t^{\ell(w)}$ introduced in 1.11; the factors will involve the degrees $d_1, \ldots, d_n$. For easy reference, we list the key ingredients needed in the proof:

(A) According to Proposition 1.11,

$$\sum_{I \subset S} (-1)^I \frac{W(t)}{W_I(t)} = t^N.$$

(B) Proposition 1.16, recalled above in 3.14, expresses $\det(w)$ as an alternating sum of class functions induced from the constant function 1 on the various parabolic subgroups of W.

(C) Lemma 3.8 allows us to compute the dimension of the space of W-invariants in a finite-dimensional W-module as the trace of the linear operator

$$\frac{1}{|W|} \sum_{w \in W} w.$$

(D) As observed in the proof of Proposition 3.8, the coefficient of t^k in the power series expansion of the product

$$\prod \frac{1}{1 - t^{d_i}}$$

is the dimension of the space R_k of homogeneous invariants of degree k.

(E) By part (c) of Proposition 3.13, the dimension of the space A_k of homogeneous alternating polynomials of degree k is equal to $\dim R_{k-N}$.

We also need an alternating sum formula based on the techniques just developed in 3.14. For this we define R_I to be the space of invariants of W_I in S; it is clearly the direct sum of its homogeneous components.

Lemma

$$\sum_{I \subset S} (-1)^I \dim(R_I)_k = \dim A_k.$$

Proof. Fix k and define a class function on W by $\chi(w) := \operatorname{Tr}_{S_k}(w)$. Denote by χ_I its restriction to W_I. From part (a) of Lemma 3.14, we get:

$$(\chi_I)^W = \chi(1^W_{W_I}),$$

whence by (B) above:

$$\sum_{I \subset S} (-1)^I (\chi_I)^W (w) = \det(w)\chi(w) \tag{30}$$

for all $w \in W$. Now average the left side of (30) over W and apply part (b) of Lemma 3.14:

$$\frac{1}{|W|} \sum_I (-1)^I \sum_{w \in W} (\chi_I)^W (w) = \sum_I (-1)^I \frac{1}{|W_I|} \sum_{z \in W_I} \chi_I(z),$$

which by (C) is the same as

$$\sum(-1)^I \dim(R_I)_k.$$

Averaging the right side of (30) over W yields:

$$\frac{1}{|W|} \sum_{w \in W} \det(w)\chi(w).$$

Thanks to (C) again, this gives the dimension of the space of W-invariants in S_k under the action $w \mapsto \det(w)w$. But the invariants under this action are just the alternating polynomials, so the right side of (30) becomes $\dim A_k$, as required. □

Theorem

$$W(t) = \prod_{i=1}^{n} \frac{t^{d_i} - 1}{t - 1}.$$

Proof. The idea is to use induction on the rank of W, taking advantage of the alternating sum formula (A) for Poincaré polynomials. Define

$$Q(t) := \prod \frac{t^{d_i} - 1}{t - 1},$$

and similarly define $Q_I(t)$ for W_I in terms of its degrees. The problem is to prove an analogue of (A) for these polynomials:

$$\sum_{I \subseteq S}(-1)^I \frac{Q(t)}{Q_I(t)} = t^N, \tag{31}$$

or, equivalently,

$$\sum(-1)^I \frac{1}{(1-t)^n Q_I(t)} = \frac{t^N}{(1-t)^n Q(t)}. \tag{32}$$

We have to compare the coefficient of t^k in the power series expansion of the rational function occurring on each side of (32). The right side equals

$$t^N \prod \frac{1}{1 - t^{d_i}},$$

so by (D) the coefficient of t^k is $\dim R_{k-N}$, which in turn equals $\dim A_k$ according to (E).

To analyze the left side of (32), consider the action of W_I on the span of Δ_I in V. Denote by $e_1, \ldots, e_I$ the degrees of basic invariants. (Since W_I fixes pointwise the orthogonal complement of this span, its degrees relative to its action on V are $e_1, \ldots, e_I, 1, \ldots, 1$.) Now

$$\frac{1}{(1-t)^n Q_I(t)} = \left(\frac{1}{(1-t)^{n-|I|}}\right) \prod \left(\frac{1}{1 - t^{e_i}}\right).$$

So the coefficient of t^k here is $\dim(R_I)_k$. Now the above lemma implies equality in (32).

By induction, for each proper parabolic subgroup W_I, we have $W_I(t)$ $= Q_I(t)$. Comparing (A) and (31), we conclude that $W(t) = Q(t)$ as required. $\square$

The theorem does not help directly to compute the degrees, but in the special case of Weyl groups it plays a key role in deriving a very effective method of computation (3.20).

Remark. The factorization of $W(t)$ in the theorem was worked out in the case of Weyl groups by Chevalley [3], as a means of simplifying his formula for the orders of finite simple groups of Lie type. Such groups have a 'Bruhat decomposition' (a double coset decomposition indexed by a Weyl group), leading to an additive expansion for the group order involving the Poincaré polynomial. To prove his formula, he relied on a related factorization of the Poincaré polynomial of a compact Lie group G having W as Weyl group; see Chevalley [1]. The Poincaré polynomial of W also plays a role here, since W parametrizes a cell decomposition of the flag manifold of G. Solomon [3] realized that the factorization of $W(t)$ is also valid for other finite reflection groups, so he developed a more elementary proof which avoids Lie groups. However, he still used a topological argument to obtain one key ingredient (our Proposition 1.16); see Solomon [4] for an algebraic version. Steinberg [5], §2, substituted a more combinatorial argument (and also generalized the theorem to cover the case of 'twisted' groups of Lie type).

3.16   Coxeter elements

Calculation of the eigenvalues of a single well-chosen element of W is enough to determine the degrees explicitly. This striking fact will be proved in 3.19, after we lay the appropriate groundwork. We assume throughout that W is *irreducible* and *essential*. It is also convenient to assume that Φ consists of unit vectors.

Enumerate a simple system Δ as $\alpha_1, \ldots, \alpha_n$, with corresponding simple reflections $s_1, \ldots, s_n$. Then $s_1 \cdots s_n$ is called a **Coxeter element** of W. Of course it depends on the choice of Δ as well as on the way Δ is numbered.

Proposition   *All Coxeter elements are conjugate in W.*

Proof. Since all simple systems are W-conjugate (1.4), it will be enough (in view of Proposition 1.2) to show that for a fixed $\Delta = \{\alpha_1, \ldots, \alpha_n\}$, the Coxeter elements resulting from different orderings of indices are conjugate.

Notice that a cyclic permutation of the indices yields a conjugate element:

$$s_n s_1 \cdots s_{n-1} = s_n(s_1 \cdots s_n)s_n, \text{ etc.}$$

It is also clear that an interchange of an adjacent *commuting* pair s_i, s_j leaves the Coxeter element unchanged. We claim that all permutations of the indices $1, 2, \ldots, n$ can be achieved by combining these two types of permutations. To get started, we can assume that n corresponds to a vertex of the Coxeter graph which is adjacent to only one other vertex. (Recall that the graph is a tree, or look at the actual list of connected Coxeter graphs in 2.4.) There is nothing to prove if $n = 1$; so we proceed by induction on n.

By induction, any permutation of $1, 2, \ldots, n-1$ alone can be achieved by a sequence of cyclic permutations and interchanges of adjacent numbers i, j for which $s_i s_j = s_j s_i$. But we have to keep track of n as well. Whenever a cyclic permutation is needed, we let n be carried along in the obvious way. The only transposition with which n can interfere involves two numbers i and j which are currently adjacent to n. But s_n commutes with at least one of s_i and s_j, say the former. So the successive interchanges of i with n and i with j are legal, and yield the desired interchange of i with j (where $s_i s_j = s_j s_i$). Thus all permutations of $1, 2, \ldots, n-1$ are achievable by cyclic permutations of all n numbers combined with interchanges of adjacent numbers corresponding to commuting reflections. In the process, n itself may get moved to an unpredictable position in the list. But because s_n commutes with all but one s_i, n can afterwards be moved to any desired position by using the permitted transpositions (and, if necessary, a further cyclic permutation). So an arbitrary permutation can be achieved by these moves. □

Since all Coxeter elements in W are conjugate, they have the same order h, which we call the **Coxeter number** of W. In some cases this is easy to compute directly. For the dihedral group $\mathcal{D}_m$, a Coxeter element is just the product of two generating reflections, hence is a rotation through $2\pi/m$, of order m. For the symmetric group $\mathcal{S}_n$, we can take s_i to be the transposition $(i, i+1)$, $1 \leq i < n$, so the corresponding Coxeter element is an n-cycle. (Thus the Coxeter number of the group of type A_{n-1} is n.) The reader might try to compute h for groups of type B_n or D_n.

The fact that Coxeter elements are all conjugate in W (hence similar in $GL(V)$) also insures that they have the same characteristic polynomial and eigenvalues. If ζ is a primitive hth root of unity in $\mathbf{C}$, these eigenvalues are of the form ζ^m, where $0 \leq m < h$. The **exponents** of W are defined to be the various m involved, written as

$$m_1 \leq m_2 \leq \ldots \leq m_n.$$

For W of type A_n, these are easily seen to be: $1, 2, ..., n$.

Our main goal will be to show that (miraculously) the degrees of W are obtained by simply increasing all the exponents by 1. This is clear already for symmetric and dihedral groups. Since the smallest degree is 2 (W being essential), we must expect that the smallest exponent is 1. In particular, w should have no nonzero fixed points. This much is easily checked:

Lemma   *A Coxeter element has no eigenvalue equal to 1. Thus the numbers $h - m_i$ are a permutation of the m_i, forcing $\sum m_i = nh/2$.*

Proof. Suppose $s_1 \cdots s_n$ fixes some λ. Then $s_2 \cdots s_n \lambda = s_1 \lambda$. The left side is congruent to λ modulo the span of $\alpha_2, \ldots, \alpha_n$, while the right side is congruent to λ modulo the span of α_1. Since the simple roots are linearly independent, this forces $s_2 \cdots s_n \lambda = \lambda = s_1 \lambda$. In particular, $(\lambda, \alpha_1) = 0$ and $s_2 \cdots s_n$ fixes λ. Repetition of the argument shows eventually that λ is orthogonal to all the basis vectors $\alpha_1, \ldots, \alpha_n$ of V. Hence $\lambda = 0$.

Since w is a real linear transformation, its nonreal eigenvalues come in complex conjugate pairs, corresponding to exponents m_i and $h - m_i$. We just saw that 1 is not an eigenvalue. The only other possible real eigenvalue is -1, which would have to be of the form $\zeta^{h/2}$. But then $m_i = h/2 = h - m_i$. So the numbers $h - m_i$ are a permutation of the m_i. This forces $\sum m_i = \sum (h - m_i)$, and in turn $\sum m_i = nh/2$. $\square$

Exercise. When W is not irreducible, Coxeter elements and the Coxeter number are defined just as above. How are they related to the Coxeter elements and numbers of the various irreducible factors of W?

3.17   Action on a plane

Consider a Coxeter element $w = s_1 \cdots s_n$. To avoid trivialities, we always assume $n \geq 2$.

How can we determine the eigenvalues of w? The most obvious approach would be to write down the matrix of w relative to the basis $\alpha_1, \ldots, \alpha_n$ of V. This is actually feasible (see Coxeter [4]), but would still leave us with the formidable task of evaluating the eigenvalues in some uniform way and then relating them to the degrees of the polynomial invariants of W. A less direct way is to study the action of w on a carefully chosen plane P in V. This leads in 3.18 to a simple formula for the Coxeter number h and also paves the way to the main theorem in 3.19.

It is a familiar fact from linear algebra that V can be decomposed into the orthogonal sum of a number of lines and planes invariant under any given orthogonal transformation (such as w). But the choice of the particular plane P we are seeking is somewhat delicate.

We claim that the simple roots can be numbered in such a way that $s_1, \ldots, s_r$ commute pairwise, as do $s_{r+1}, \ldots, s_n$, for some $r < n$. As in 3.16, this relies on the fact that the Coxeter graph is a tree, so that some vertex is connected to only one other vertex. By removing this vertex and using induction, our claim follows at once. We fix this numbering of simple roots, and the corresponding Coxeter element w.

Now we have a partition of $K := \{1, 2, \ldots, n\}$ as $K = I \cup J$, where $I := \{1, \ldots, r\}$ and $J := \{r+1, \ldots, n\}$. This induces a factorization $w = yz$, where $y := s_1 \cdots s_r$ and $z := s_{r+1} \cdots s_n$. The choice of numbering implies that each of y and z has order 2 (as a product of commuting reflections).

Next define $\omega_1, \ldots, \omega_n$ to be the dual basis of $\alpha_1, \ldots, \alpha_n$ in V. For reasons of dimension, the span Y of $\omega_{r+1}, \ldots, \omega_n$ has orthogonal complement $Y^\perp$ spanned by $\alpha_1, \ldots, \alpha_r$. Similarly, the span Z of $\omega_1, \ldots, \omega_r$ has orthogonal complement spanned by $\alpha_{r+1}, \ldots, \alpha_n$. If H_i is the hyperplane orthogonal to α_i, it is clear that $Y \subset Y' := H_1 \cap \ldots \cap H_r$ and that $Z \subset Z' := H_{r+1} \cap \ldots \cap H_n$. At the same time, $Y' \cap Z' = 0$ (since only 0 is orthogonal to all α_i). Since $V = Y \oplus Z$, we conclude that $Y = Y'$ and $Z = Z'$. Obviously y fixes Y pointwise and acts on $Y^\perp$ as -1. Similarly, z fixes Z pointwise and acts as -1 on $Z^\perp$.

The final ingredient we need is the (positive definite) matrix A of the bilinear form associated with the Coxeter graph: $A = (a_{ij})$, where $a_{ij} = (\alpha_i, \alpha_j)$. In our current set-up, A is the matrix (relative to the dual basis) of the linear operator on V sending each ω_i to the corresponding α_i, since $\alpha_j = \sum_i a_{ij} \omega_i$. Now we can appeal to the general fact proved in Proposition 2.6 (in the case of an indecomposable positive definite symmetric matrix with nonpositive off-diagonal entries): A has a positive eigenvalue c with a corresponding eigenvector $(c_1, \ldots, c_n)$ in $\mathbf{R}^n$ having all $c_i > 0$.

Set

$$\lambda := \sum_{i \in I} c_i \omega_i, \quad \mu := \sum_{j \in J} c_j \omega_j.$$

Thus $\lambda \in Z$ and $\mu \in Y$. We want to show that w stabilizes the plane P spanned by the lines $L := \mathbf{R}\lambda$ and $M := \mathbf{R}\mu$. By choice of the eigenvalue c (and the description of A as a linear operator), we have

$$\sum_{k \in K} c_k \alpha_k = \sum_{k \in K} cc_k \omega_k.$$

Taking the inner product of each side with $\alpha_i (i \in I)$, we get

$$c_i + \sum_{j \in J} c_j a_{ij} = cc_i,$$

since $(\alpha_i, \alpha_l) = 0$ for all $i \neq l$ in I. We use this to calculate as follows (with sums over $i \in I, j \in J, k \in K$ respectively):

$$
\begin{aligned}
(c-1)\lambda &= (c-1) \sum_i c_i \omega_i \\
&= \sum_i \left(\sum_j c_j a_{ij} \right) \omega_i \\
&= \sum_j c_j \left(\sum_i a_{ij} \omega_i \right) \\
&= \sum_j c_j \left(-\omega_j + \sum_k a_{kj} \omega_k \right) \\
&= -\sum_j c_j \omega_j + \sum_j c_j \alpha_j \\
&= -\mu + \nu,
\end{aligned}
$$

where ν is orthogonal to $\omega_1, \ldots, \omega_r$ (hence to Z). In the fourth step we were able to sum over all of K by using the fact that $a_{kj} = 0$ for $k \in J$ (unless $k = j$, in which case $a_{jj} = 1$). In the fifth step we used again the fact that $\sum_k a_{kj} \omega_k = \alpha_j$.

The calculation shows that the linear combination $\nu = (c-1)\lambda + \mu$ is orthogonal to Z (hence is sent to its negative by z), while λ lies in Z (hence is fixed by z). It follows that z stabilizes the span of these two vectors, which is just P. A similar argument shows that y stabilizes P. Moreover, y (resp. z) acts on P as a reflection with fixed line M (resp. L). Thus w acts on P as a rotation.

What is the order of w, acting on P? Note that λ and μ lie in the fundamental domain $\bar{C}$ of W defined in 1.12, since C consists of the positive linear combinations of the ω_i, and $\bar{C}$ of the nonnegative combinations. Clearly $P \cap C$ consists of all $a\lambda + b\mu$ with $a, b > 0$; in particular, P meets C. If w^t fixes P pointwise, then it fixes some element of $C \cap P$, so by 1.12 $w^t = 1$. It follows that w has order precisely h on P (so that w acts as a rotation through $2\pi/h$).

As a corollary of this discussion, we see that the primitive hth root of unity ζ actually does occur as an eigenvalue of w (since it occurs already as an eigenvalue of the plane rotation through $2\pi/h$). The inverse (complex conjugate) of this eigenvalue also occurs, of course. So $m_1 = 1$, as our preview of the main theorem led us to expect, and $m_n = h - 1$. (This is true even when $n = 1$, since then $h = 2$.)

Proposition *If the exponents of W are listed as $m_1 \leq m_2 \leq \ldots \leq m_n$, we have $m_1 = 1$ and $m_n = h - 1$.* $\square$

Exercise. Is the partition $K = I \cup J$ unique? Describe such a partition for each individual Coxeter graph.

3.18 The Coxeter number

Now we can exploit the way w acts on the plane P to derive a simple formula for h. The promised comparison between exponents and degrees (to be carried out in the next section) suggests what the formula should look like. Recall from Lemma 3.16 that the sum of the exponents is $nh/2$. On the other hand, Theorem 3.9 implies that the sum of the numbers $d_i - 1$ is N. The two answers do agree:

Proposition *The Coxeter number $h = 2N/n$, where N is the number of positive roots.*

Proof. We may assume $n > 1$. The idea is to see how the N reflecting hyperplanes $H_\alpha (\alpha \in \Phi^+)$ intersect the plane P. Since P contains points of C, no H_α contains P, so each such intersection is a line.

From our previous description of P, we see that rotating the lines L and M by powers of w produces a total of h lines, all in one orbit under w if h is odd, but in separate orbits containing L and M respectively if h is even. All points of P not in these lines are obtained by rotating points of C. In particular, each H_α must intersect P in one of the indicated lines.

Since $L \subset Z = H_{r+1} \cap \cdots \cap H_n$, it is clear that these $n-r$ hyperplanes intersect P in L. We claim no other H_α can do so. Suppose $H_\alpha \cap P = L$, and write $\alpha = \sum r_k \alpha_k$ (with $r_k \geq 0$). To say that $\lambda \in H_\alpha$ is to say that

$$0 = (\lambda, \alpha) = \left(\sum_{i \in I} c_i \omega_i, \sum_{k \in K} r_k \alpha_k \right) = \sum_{i \in I} c_i r_i.$$

Since all $c_i > 0$, this forces $r_i = 0 \ (i \leq r)$. But the choice of J insures that the only positive roots obtainable as linear combinations of the $\alpha_j \ (j \in J)$ are the α_j themselves (see Proposition 1.10).

Similarly, the only hyperplanes intersecting P in M are $H_1, \ldots, H_r$.

If h is even, it is clear that the number of hyperplanes H_α intersecting P in each of the $h/2$ distinct lines $w^t L$ is $n-r$, giving a total of $(n-r)h/2$. The remaining $h/2$ distinct lines $w^t M$ account for an additional $rh/2$ hyperplanes, giving a grand total of $N = nh/2$.

If h is odd, the number of H_α intersecting P in each of the h lines $w^t L$ is $(n-r)$, giving a total of $N = (n-r)h$. But the h lines are also of the form $w^t M$, giving a total of $N = rh$. This forces $r = n/2$, so again $N = nh/2$. $\square$

Recalling the discussion at the start of this section, we conclude:

Corollary $\sum_{i=1}^{n} m_i = N = \sum_{i=1}^{n} (d_i - 1).\square$

From the proposition we see that $h = 2$ can occur only if $N = 1 = n$, which happens only for W of type A_1. (In all other cases ζ is *nonreal*, a fact which will be needed in the following section.) By comparing Table 2 in 2.11, one quickly finds all values of h as listed in the table below.

A_n	B_n	D_n	E_6	E_7	E_8	F_4	G_2	H_3	H_4	$I_2(m)$
$n(n+1)/2$	n^2	$n(n-1)$	36	63	120	24	6	15	60	m
$n+1$	$2n$	$2(n-1)$	12	18	30	12	6	10	30	m

Table 2: Number of positive roots and Coxeter number

Example. The group of type H_3 has three exponents, whose sum must be $N = 15$ by the corollary above. Proposition 3.17 insures that 1 and $h - 1 = 9$ are exponents, so the remaining one must be $5 = h - 5$. (What can be said at this point about groups of type H_4 and F_4?)

3.19 Eigenvalues of Coxeter elements

At last we are ready to relate the exponents and the degrees of W. In order to bring the eigenvalues of the Coxeter element w into the picture, we have to complexify the situation. Embed V in its complexification $V_{\mathbf{C}}$, and view the inner product on V as the restriction of a (unitary) inner product on $V_{\mathbf{C}}$. With respect to an ordered basis of V, we are just embedding $\mathbf{R}^n$ in $\mathbf{C}^n$. Similarly, the ring S of polynomial functions, identified with $\mathbf{R}[x_1, \ldots, x_n]$, has the ring of complex polynomials as its complexification.

In the complex setting, we make an important observation about eigenvectors of w. Still excluding the group of rank 1, we found as a consequence of Proposition 3.18 that the eigenvalue ζ (a primitive hth root of unity) is nonreal. So the plane P contains no eigenvector, though its complexification $P_{\mathbf{C}}$ does contain an eigenvector κ along with the distinct complex conjugate vector $\bar{\kappa}$; these vectors span $P_{\mathbf{C}}$. We claim that κ cannot be orthogonal to any root α. From $(\kappa, \alpha) = 0$ we would get $(\bar{\kappa}, \alpha) = 0$ as well, forcing $P_{\mathbf{C}}$ to be orthogonal to α. This is impossible, because P contains points of C and therefore does not lie in H_α.

Let $\lambda_1, \ldots, \lambda_n$ be a basis of $V_{\mathbf{C}}$ consisting of eigenvectors for w, relative to the eigenvalues ζ^{m_i}, and denote by $y_1, \ldots, y_n$ the corresponding coordinate functions, which generate $S_{\mathbf{C}}$. In view of the way W acts on polynomial functions, we have

$$w \cdot y_i = \zeta^{-m_i} y_i.$$

Now let $f_1, \ldots, f_n$ be a basic set of invariants in S (of degrees $d_1, \ldots, d_n$), and express them as polynomials (with complex coefficients) in the y_i. Since the change of variables is linear (over $\mathbf{C}$), the original Jacobian determinant $J = J(f_1, \ldots, f_n)$ described in 3.10 is altered only by a nonzero scalar factor, which we can ignore. Recall from 3.13 that J can be factored into the product of linear polynomials whose zero sets are the N root hyperplanes $H_\alpha, \alpha \in \Phi^+$. Such a factorization still remains after the change of coordinates from x_i to y_i.

Theorem *If $m_1, \ldots, m_n$ are the exponents of W, then the degrees of W are $m_1 + 1, \ldots, m_n + 1$. Therefore $|W| = \prod(m_i + 1)$.*

Proof. As before, we may assume $n > 1$. We choose the numbering of the eigenvectors of w so that λ_1 has eigenvalue ζ $(m_1 = 1)$. As observed above, this eigenvector lies in no hyperplane H_α, so $J(1, 0, \ldots, 0) \neq 0$. Thus at least one of the $n!$ products involved in the determinant is nonzero at this point. By renumbering the invariants $f_1, \ldots, f_n$ suitably, we can therefore assume that all $\partial f_i / \partial y_i$ are nonzero at $(1, 0, \ldots, 0)$. This just means that

$$\frac{\partial f_i}{\partial y_i} = a_i y_1^{d_i - 1} + \text{ terms involving } y_2, \ldots, y_n,$$

with $a_i \neq 0$. In turn,

$$f_i = a_i y_1^{d_i - 1} y_i + \text{ terms involving other monomials.}$$

Now apply w to this equation, bearing in mind that $w \cdot y_i = \zeta^{-m_i} y_i$:

$$f_i = w \cdot f_i = a_i \zeta^{1 - d_i - m_i} y_1^{d_i - 1} y_i + \text{ terms involving other monomials.}$$

This forces

$$\zeta^{1 - d_i - m_i} = 1,$$

whence

$$d_i - 1 \equiv -m_i \equiv h - m_i \pmod{h}.$$

By Lemma 3.16, the numbers $h - m_i$ are a permutation of the numbers m_i and, by Corollary 3.18, their sum equals the sum of the $d_i - 1$. From $d_i - 1 \equiv h - m_i \pmod{h}$ and $0 < m_i < h$ we finally get $d_i - 1 = h - m_i$. This shows that the $d_i - 1$ are equal to the exponents, as required. Finally, Theorem 3.9 shows that $|W| = \prod(m_i + 1)$. $\square$

Example. Now we can easily determine the exponents (and degrees) of the group of type H_4, for which $N = 60, h = 30, |W| = 14\,400$. Since 1 and 29 must occur, the other exponents a, b must add up to 30, while $(a+1)(b+1) = 14\,400/60 = 240$. The solutions of the resulting quadratic equation are 11 and 19. Thus the degrees are 2, 12, 20, 30, in agreement with Table 1. (The reader should do a similar calculation for the group of type F_4.)

Exercise 1. In the list of exponents of W, show that 1 and $h - 1$ occur only once.

Corollary   *The scalar transformation* $-1 \in \mathrm{GL}(V)$ *lies in W if and only if all exponents of W are odd (or if and only if all degrees are even). In this case, h must be even and, for any Coxeter element w, we have* $-1 = w^{h/2}$.

Proof. The transformation -1 on V induces in a canonical way an automorphism of the symmetric algebra S, acting on S_d as $(-1)^d$. If $-1 \in W$, it follows that no W-invariant polynomial of odd degree can exist. So the degrees are all even, and (by the theorem) the exponents are all odd.

Conversely, let all exponents be odd. Since $h - 1$ occurs as an exponent, h must be even. Consider $z := w^{h/2}$, where w is a Coxeter element. Then z acts on eigenvectors of w by the scalars

$$\zeta^{m_i h/2}$$

which are square roots of 1 but not equal to 1 (since m_i is odd). Thus $z = -1$. In particular, $-1 \in W$. $\square$

A glance at Table 1 in 3.7 shows in which irreducible cases we have $-1 \in W$. The exceptions are types A_n $(n \geq 2), D_n$ $(n$ odd$), E_6, I_2(m)$ $(m$ odd$)$.

Exercise 2. If h is even and $w = s_1 \cdots s_n$ is a Coxeter element, set $z := w^{h/2}$. Show that z is the longest element $w_\circ$ of W (relative to Δ), so $\ell(z) = N$, and exhibit a reduced expression for z. [Use 3.17 to show that z maps C to $-C$.]

3.20 Exponents and degrees of Weyl groups

In this final section we explore some special features of crystallographic reflection groups (Weyl groups). The first result gives a surprisingly easy way to determine many of the exponents.

Proposition   *Let W be an irreducible Weyl group, with Coxeter number h. If $1 \leq m \leq h-1$ and m is relatively prime to h, then m is an exponent of W.*

Proof. Since W stabilizes its root lattice $L(\Phi)$, the matrix of a Coxeter element w relative to a basis of V consisting of simple roots will have entries in $\mathbf{Z}$. In turn, the characteristic polynomial has integral coefficients. According to Proposition 3.17, a primitive hth root of unity ζ occurs as an eigenvalue of w. The distinct primitive hth roots of unity are precisely the powers ζ^m, with m relatively prime to h and $1 \le m \le h - 1$. Now it is well-known that the corresponding cyclotomic polynomial (having all primitive hth roots of unity as roots) is irreducible over $\mathbf{Z}$, so it must divide the characteristic polynomial of w, with which it has a greatest common divisor of positive degree. $\square$

This makes it easy to complete the determination of exponents (and hence degrees) for the exceptional Weyl groups. In the case of the group of type F_4, we could do the computation earlier by combining several facts. But our new criterion is much quicker, since 1, 5, 7, 11 are all relatively prime to 12.

In the case of the groups of type E_n, none of the previous techniques is adequate to complete the list, unless we are willing to analyze the matrices of Coxeter elements. But now the task is easy.

For the group of type E_8, with $h = 30$, there happen to be precisely eight values of m between 1 and 29 which are relatively prime to 30:

$$1, 7, 11, 13, 17, 19, 23, 29$$

So these must be the exponents! For E_7, with $h = 18$, we find only six of the seven exponents this way: $1, 5, 7, 11, 13, 17$. But then the missing m must equal $h - m$, forcing $m = h/2 = 9$ (cf. Lemma 3.16). For E_6, with $h = 12$, the numbers $1, 5, 7, 11$ must all be exponents. The remaining ones are forced to be 4 and 8, since the sum of exponents is $N = 36$ and the product of degrees is $|W| = 2^7 3^4 5$.

Exercise 1. Does the proposition remain valid for any of the non-crystallographic groups?

Exercise 2. The exponents whose existence is guaranteed by the proposition occur with multiplicity 1.

We conclude by stating a theorem which provides a completely different approach to the computation of exponents (hence degrees), based on the 'height' of roots in a crystallographic root system. Because the theorem is valid only for Weyl groups, and because its proof requires some ideas most naturally developed in the context of representations of Lie groups, we shall be content with a rough outline of the proof. (See Carter [1], 10.1–10.2, for complete details.)

Fix a set Δ of simple roots in the (crystallographic) root system Φ, and define the **height** $h(\alpha)$ of a root α to be the sum of the coefficients of α when expressed as a linear combination of Δ. Then let k_i be the number of positive roots of height i, for each $i > 0$. For example, $k_1 = n$, since only the n simple roots have height 1. By examining each root system, one finds (somewhat surprisingly) that in all cases:

$$k_1 \geq k_2 \geq \cdots$$

Thus we have a partition of the number N of positive roots, written in standard nonincreasing order. Such a partition has a dual partition $\ell_1 \geq \ell_2 \geq \cdots$, where ℓ_j is defined to be the number of $k_i \geq j$. (This corresponds to transposing the 'Young diagram', a topheavy array of boxes with k_i boxes in the ith row.) Note that the dual partition has n parts, since $k_1 = n$. Now we can state:

Theorem *Let W be a Weyl group, with exponents $1 = m_1 \leq \cdots \leq m_n = h - 1$. Written in reverse order $m_n \geq \cdots \geq m_1$ as a partition of N, its dual partition is $n = k_1 \geq \cdots \geq k_{h-1} = 1$, where k_i is the number of positive roots of height i. (In particular, the highest root $\tilde{\alpha}$ has height $h - 1$.)*

Before sketching the proof, we should emphasize that a rigorous case-by-case proof is possible, based on close study of the individual root systems. Indeed, the theorem was first verified empirically in this way. For the exceptional Weyl groups, one can consult the explicit lists of positive roots (arranged by height) in Springer [1]. But naturally one would prefer a general proof which explains what is going on.

In view of 3.19, the theorem can be derived by factoring the Poincaré polynomial $W(t) = \sum_{w \in W} t^{\ell(w)}$ in a new way and comparing the factorization obtained previously in 3.15 (for an arbitrary finite reflection group W):

$$W(t) = \prod_{i=1}^{n} \frac{t^{d_i} - 1}{t - 1}. \tag{33}$$

The new factorization makes sense only for Weyl groups:

$$W(t) = \prod_{\alpha > 0} \frac{t^{h(\alpha)+1} - 1}{t^{h(\alpha)} - 1}. \tag{34}$$

The product in (34) is taken over all positive roots, but permits a considerable amount of cancellation (when N is large compared with n). The idea is already clear for the root system of type B_2. Say α is the long simple root, β the short simple root. The other positive roots are $\alpha + \beta$ and $\alpha + 2\beta$, so $(k_1, k_2, k_3) = (2, 1, 1)$, and the dual partition is $(3, 1)$. The right side of (34) looks like

$$\frac{t^2 - 1}{t - 1} \cdot \frac{t^2 - 1}{t - 1} \cdot \frac{t^3 - 1}{t^2 - 1} \cdot \frac{t^4 - 1}{t^3 - 1}.$$

After cancelling some numerators with subsequent denominators, this becomes the right side of (33):

$$\frac{(t^2 - 1)(t^4 - 1)}{(t - 1)^2}.$$

The identity (34) is actually a specialization of a more sophisticated-looking identity, which requires the introduction of the group algebra B (over $\mathbf{Q}$) of the abelian group $L(\Phi)$ (the root lattice). To write the group operation multiplicatively, use symbols $e(\lambda)$ in bijection with elements $\lambda \in L(\Phi)$, with $e(\lambda)e(\mu) = e(\lambda + \mu)$. Then B consists of the formal $\mathbf{Q}$-linear combinations of the $e(\lambda)$, and is easily seen to be an integral domain. We can also work in the polynomial ring $B[t]$ or its fraction field. With this notation, the more general identity reads:

$$W(t) = \sum_{w \in W} \left(\prod_{\alpha > 0} \frac{1 - te(-w\alpha)}{1 - e(-w\alpha)} \right). \tag{35}$$

To specialize this to (34), define a homomorphism $\psi : B[t] \rightarrow \mathbf{Q}[t, t^{-1}]$, viewing the latter as the group algebra of the infinite cyclic group with generator t. Here ψ sends t to t and sends a typical $e(\alpha)$ to $t^{-h(\alpha)}$. The fractions occurring in (35) actually represent elements of $B[t]$, so we can apply ψ (leaving $W(t)$ unchanged). In the sum over W, all terms for which $w \neq 1$ get sent to 0, since there exists $\alpha > 0$ for which $w\alpha$ has height -1 (i.e., $w^{-1} \neq 1$ sends some element of $-\Delta$ to Φ^+).

Where does (35) come from? One needs to work with the weight lattice $\hat{L}(\Phi)$ and its rational group algebra $\hat{B}$ (which includes B). The particular weight $\rho := \frac{1}{2} \sum_{\alpha > 0} \alpha$ occurs very often in Lie theory. There is a notion of 'alternating' element for the action of W on $\hat{B}$, like that in 3.13. The operator $\theta := \sum_{w \in W} \det(w) w$ maps $\hat{B}$ into its alternating elements. In the framework of Weyl's character formula, one finds the identity:

$$\theta(e(\rho)) = e(-\rho) \prod_{\alpha > 0} (e(\alpha) - 1).$$

Because the right side is an alternating element, we get for any $w \in W$

$$\prod_{\alpha > 0} (1 - e(-w\alpha)) = e(-w\rho) \det(w)\theta(e(\rho)),$$

to be substituted for the denominator in (35). Expand the product in the numerator into a sum (over subsets of Φ^+) and interchange the order of summation. After some delicate manipulation, one finds that $\theta(e(\rho))$ also appears in the numerator, and can therefore be cancelled. Eventually just $W(t)$ is left.

Notes

As indicated by the length of this chapter, the invariant theory of finite reflection groups is a rich and highly applicable subject. Useful accounts are found in many places, including Bourbaki [1], V, §5–6, Carter [1], Chapters 9–10, Flatto [3], Hiller [3], Chapter II, Springer [3], Steinberg [4], §9.

(3.5) Case-by-case treatments of rings of invariants appear in Coxeter [4] and Shephard–Todd [1], but Chevalley [2] gave the first unified approach.

(3.11) See Shephard–Todd [1]. A version valid for arbitrary fields is given by Kac–Watanabe [1].

(3.12) Explicit (or algorithmic) descriptions of basic invariants appear in many places, e.g., Coxeter [4], Flatto [1][3], Ignatenko [1], Mehta [1], Saito–Yano–Sekiguchi [1], Sekiguchi–Yano [1][2].

(3.13) According to Coxeter [4], 6.2, this factorization of the Jacobian was conjectured by J.A. Todd and proved by G. Racah.

(3.16)–(3.19) After initial observations by W. Killing and others, these ideas were systematized by Coxeter [2][4], whose proofs sometimes involve case-by-case verifications. He remarks ([4], p. 765), 'Having computed the m's several years earlier, I recognized them in the Poincaré polynomials while listening to Chevalley's address at the International Congress in 1950.' Theorem 3.19 explains neatly a symmetry in the exponents observed by Chevalley [1]. Coleman [1] gave a more unified proof of the theorem, using however Coxeter's empirical observation that $h = 2N/n$; then Steinberg [1] provided a uniform proof of the latter, which we have followed. (At about the same time Kostant [1] gave a Lie-theoretic proof of this formula.) For accounts similar to ours, see Carter [1], Chapter 10, Bourbaki [1], V, §6. See also (8.4) below for generalizations to other Coxeter groups.

(3.20) The proposition appears in Bourbaki [1], VI, 1.11, along with other interesting facts about the Coxeter number. (Note that the proof of Prop. 33 was revised in the later edition.) The theorem was apparently first discovered by A. Shapiro (unpublished), then proved uniformly in a Lie algebra context by Kostant [1]. Macdonald [2] instead derives the theorem from the identity (35), of which he proves a more general version. Recently Akyildiz–Carrell [1] have placed the theorem itself in a more general geometric setting.

Chapter 4

Affine reflection groups

In this chapter we describe a class of infinite groups generated by affine reflections in euclidean space, which are intimately related to Weyl groups and which turn out to have a presentation like that of finite reflection groups (1.9). This will help to motivate the general study of Coxeter groups in Part II. For the most part we follow Iwahori–Matsumoto [1]. (See also Bourbaki [1], VI, §2.)

Throughout this chapter W denotes a Weyl group (a finite crystallographic reflection group), acting on the euclidean space V, as described in 2.9. The corresponding (crystallographic) root system is denoted Φ. We also need the set $\Phi^\vee$ of coroots $\alpha^\vee = 2\alpha/(\alpha, \alpha)$, which is a root system in V in its own right with Weyl group W.

4.1   Affine reflections

We want to consider not just orthogonal reflections (leaving the origin in V fixed), but also **affine reflections** relative to hyperplanes which do not necessarily pass through the origin. To this end we introduce the **affine group** $\mathrm{Aff}(V)$, which is the semidirect product of $\mathrm{GL}(V)$ and the group of translations by elements of V. To each $\lambda \in V$ we associate the translation $t(\lambda)$, which sends $\mu \in V$ to $\mu + \lambda$. Then we see immediately that, for any $g \in \mathrm{GL}(V)$ and $\lambda \in V$,

$$gt(\lambda)g^{-1} = t(g\lambda),$$

showing that the group of translations is indeed normalized by $\mathrm{GL}(V)$.

For each root α and each integer k, define an affine hyperplane

$$H_{\alpha,k} := \{\lambda \in V | (\lambda, \alpha) = k\}.$$

Note that $H_{\alpha,k} = H_{-\alpha,-k}$ and that $H_{\alpha,0}$ coincides with the reflecting hyperplane H_α. Note too that $H_{\alpha,k}$ can be obtained by translating H_α

by $\frac{k}{2}\alpha^\vee$. Define the corresponding affine reflection as follows:

$$s_{\alpha,k}(\lambda) := \lambda - ((\lambda,\alpha) - k)\alpha^\vee.$$

This is geometrically correct, because it fixes $H_{\alpha,k}$ pointwise and sends the 0 vector to $k\alpha^\vee$. We can also write $s_{\alpha,k}$ as $t(k\alpha^\vee)s_\alpha$. In particular, $s_{\alpha,0} = s_\alpha$.

Denote by $\mathcal{H}$ the collection of all hyperplanes $H_{\alpha,k}$ ($\alpha \in \Phi, k \in \mathbf{Z}$). The following proposition (the proof of which is an immediate calculation) shows that the elements of $\mathcal{H}$ are permuted in a natural way by W as well as by certain translations in Aff(V).

Proposition

(a) *If $w \in W$, then $wH_{\alpha,k} = H_{w\alpha,k}$ and $ws_{\alpha,k}w^{-1} = s_{w\alpha,k}$.*

(b) *If $\lambda \in V$ satisfies $(\lambda,\alpha) \in \mathbf{Z}$ for all roots α, then $t(\lambda)H_{\alpha,k} = H_{\alpha,k+(\lambda,\alpha)}$ and $t(\lambda)s_{\alpha,k}t(-\lambda) = s_{\alpha,k+(\lambda,\alpha)}$.* □

4.2   Affine Weyl groups

We define the **affine Weyl group** W_a to be the subgroup of Aff(V) generated by all affine reflections $s_{\alpha,k}$, where $\alpha \in \Phi, k \in \mathbf{Z}$.

Example. If $|W| = 2$, the corresponding group W_a is generated by s_α together with $s_{\alpha,1}$ subject only to the obvious relations (the square of each reflection is 1). This group is called the **infinite dihedral group**, denoted $\mathcal{D}_\infty$. To emphasize the connection with the group of type A_1, we say that W_a is of type $\widetilde{A_1}$. (Similar notation is used for other types.)

We can make the structure of W_a more transparent. Recall from 2.9 the **root lattice** $L(\Phi)$ (the $\mathbf{Z}$-span of Φ) and the **weight lattice**

$$\hat{L}(\Phi) = \{\lambda \in V | (\lambda,\alpha^\vee) \in \mathbf{Z} \text{ for all } \alpha \in \Phi\}.$$

Similarly, we obtain lattices associated with the root system $\Phi^\vee$. Set $L := L(\Phi^\vee)$ and $\hat{L} := \hat{L}(\Phi^\vee)$, the latter characterized by:

$$\hat{L} = \{\lambda \in V | (\lambda,\alpha) \in \mathbf{Z} \text{ for all } \alpha \in \Phi.\}$$

(This is the condition appearing in part (b) of Proposition 4.1.) W stabilizes each of these lattices, which we identify with the corresponding translation groups in Aff(V).

Proposition W_a *is the semidirect product of W and the translation group corresponding to the coroot lattice $L = L(\Phi^\vee)$.*

Proof. W normalizes L, and obviously they have trivial intersection; denote their semidirect product by W'. We saw in 4.1 that $s_{\alpha,k} =$

$t(k\alpha^\vee)s_\alpha$, so the generators of W_a all lie in W'. The same equation shows that $t(k\alpha^\vee) = s_{\alpha,k}s_\alpha$ lies in W_a, so both L and W are included in W_a. $\square$

Since the translation group corresponding to $\hat{L}$ is also normalized by W, we can form the semidirect product (called $\widehat{W_a}$), which contains W_a as a normal subgroup of finite index. Indeed, $\widehat{W_a}/W_a$ is isomorphic to $\hat{L}/L$. As a matter of notation, we shall use letters such as w to denote arbitrary elements of $\widehat{W_a}$ throughout this chapter.

From Proposition 4.1 we deduce that $\widehat{W_a}$ permutes the hyperplanes in $\mathcal{H}$. More precisely:

Corollary *If $w \in \widehat{W_a}$ and $H_{\alpha,k} \in \mathcal{H}$, then $wH_{\alpha,k} = H_{\beta,l}$ for some $\beta \in \Phi$, $l \in \mathbf{Z}$, and thus $w s_{\alpha,k} w^{-1} = s_{\beta,l}$.* $\square$

4.3 Alcoves

To study how the groups $\widehat{W_a}$ and W_a permute the hyperplanes in $\mathcal{H}$, we examine how they permute the collection $\mathcal{A}$ of connected components of $V^\circ := V \setminus \bigcup_{H \in \mathcal{H}} H$. Each element of $\mathcal{A}$ is called an **alcove**. It is clear (since elements of $\mathrm{Aff}(V)$ act as homeomorphisms) that $\widehat{W_a}$ does permute $\mathcal{A}$.

What do alcoves look like? Notice first that V° is *open* in V. Given $\lambda \in V^\circ$, for each root α there is some $k \in \mathbf{Z}$ such that λ lies between $H_{\alpha,k}$ and $H_{\alpha,k+1}$, so we can find an open neighborhood U_α of λ meeting no α-hyperplane. Intersecting these neighborhoods for all roots α yields an open neighborhood of λ in V°. Since V° is open, its connected components are also open.

From now on, we fix a set Δ of simple roots in Φ. We assume moreover that Φ is *irreducible*. It is convenient to single out one particular alcove:
$$A_\circ := \{\lambda \in V \mid 0 < (\lambda,\alpha) < 1 \text{ for all } \alpha \in \Phi^+\}.$$

This really is an alcove. On the one hand, it is clearly included in V°. On the other hand, it is convex (hence connected), but any element outside $A_\circ$ is separated from it by one of the hyperplanes H_α or $H_{\alpha,1}$; so $A_\circ$ is a connected component of V°. In general, an alcove is defined by a set of inequalities (some of which may be redundant) of the form:
$$k_\alpha < (\lambda,\alpha) < k_\alpha + 1, \quad \alpha \in \Phi^+.$$

Example. The alcoves associated with W_a when W has rank 2 are triangles having angles π/k, π/l, π/m, where $(k,l,m) = (3,3,3)$, $(2,4,4)$, $(2,3,6)$, in the respective cases $\widetilde{A_2}$, $\widetilde{B_2}$, $\widetilde{G_2}$.

Exercise. An alcove A consists of all $\lambda \in V$ satisfying strict inequalities $k_\alpha < (\lambda, \alpha) < k_\alpha + 1$, where α runs over Φ^+ and $k_\alpha \in \mathbf{Z}$. Its **upper closure** consists of those λ satisfying the inequalities obtained by replacing the second $<$ by $\leq$ in each case. Prove that each $\lambda \in V$ lies in the upper closure of a unique alcove.

Since Φ is irreducible, there is a unique highest root $\tilde{\alpha}$ (which is long if there are two root lengths), having the property that, for all positive roots α, $\tilde{\alpha} - \alpha$ is a sum of simple roots (2.9). We claim that

$$A_\circ = \{\lambda \in V \,|\, 0 < (\lambda, \alpha) \text{ for all } \alpha \in \Delta, (\lambda, \tilde{\alpha}) < 1\}.$$

It is obvious that $A_\circ$ is included in this set. On the other hand, if λ satisfies the indicated inequalities, note that also $(\lambda, \alpha) > 0$ for all positive α. Since $\tilde{\alpha} - \alpha$ is a sum of simple roots, $(\lambda, \tilde{\alpha} - \alpha) \geq 0$, so $(\lambda, \alpha) \leq (\lambda, \tilde{\alpha}) < 1$ and thus $\lambda \in A_\circ$.

This description shows that $A_\circ$ is simply an intersection of open half-spaces. Moreover, it is a euclidean simplex (whereas if Φ has a number of irreducible components, $A_\circ$ will be a product of simplexes). (*Question:* What are the vertices of $A_\circ$?)

Define the **walls** of $A_\circ$ to be the hyperplanes H_α, $\alpha \in \Delta$ and $H_{\tilde{\alpha},1}$, and define S_a to be the corresponding set of reflections:

$$S_a := \{s_\alpha, \alpha \in \Delta\} \cup \{s_{\tilde{\alpha},1}\}.$$

The walls of $wA_\circ$ can then be defined to be the images of these hyperplanes under w for any $w \in W_a$. As soon as we show that W_a acts transitively on $\mathcal{A}$, we will have well-defined walls for each alcove.

Proposition *The group W_a permutes the collection $\mathcal{A}$ of all alcoves transitively, and is generated by the set S_a of reflections with respect to the walls of the alcove $A_\circ$.*

Proof. Let W' be the subgroup of W_a generated by S_a. We first prove that W' permutes $\mathcal{A}$ transitively. For this it is enough to show: for any alcove A, there exists $w \in W'$ for which $wA = A_\circ$. Fix any two elements $\lambda \in A_\circ, \mu \in A$. It is clear that the orbit of μ under the translation group $L = L(\Phi^\vee)$ is a discrete subset of V. Since W_a is an extension of this lattice by a finite group, the W_a-orbit (and *a fortiori* the W'-orbit) of μ is also discrete in V. So this orbit contains an element $\nu = w\mu$ of smallest possible distance from λ. If we can show that $\nu \in A_\circ$, it will follow that $wA \cap A_\circ \neq \emptyset$, and thus $wA = A_\circ$.

Suppose instead that $\nu \notin A_\circ$. Then λ and ν must lie in different half-spaces relative to some wall H of $A_\circ$. Let s be the corresponding reflection (so $s \in W'$). Consider the trapezoid in V (which H bisects) having vertices $\nu, s\nu, s\lambda, \lambda$. It is an elementary geometric fact (proved using the Law of Cosines, for example) that the length of a diagonal is greater than the common length of the two nonparallel sides of the trapezoid. This translates into the inequality: $\|s\nu - \lambda\| < \|\nu - \lambda\|$. Because $s\nu$ is in the W'-orbit of μ, this contradicts the choice of ν.

We have shown that W' permutes $\mathcal{A}$ transitively. In particular, each alcove has well-defined 'walls' (the images of the walls of A_o), and every hyperplane $H_{\alpha,k}$ occurs as a wall of two or more alcoves. To show that $W' = W_a$ we just have to see that each $s_{\alpha,k}$ lies in W'. Let A be any alcove having $H_{\alpha,k}$ as a wall, and find $w \in W'$ for which $wA = A_o$. Then $wH_{\alpha,k}$ coincides with one of the walls H of A_o, whose corresponding reflection s lies in W'. By Corollary 4.2, $ws_{\alpha,k}w^{-1} = s$, forcing $s_{\alpha,k} \in W'$ as desired. $\square$

Since S_a generates W_a, it is natural (imitating the procedure in Chapter 1) to define the **length** $\ell(w)$ of an element $w \in W_a$ to be the smallest r for which w is a product of r elements of S_a; such an expression is called **reduced**. Our next task is to give a geometric characterization of the length function.

4.4 Counting hyperplanes

Given a hyperplane $H = H_{\alpha,k} \in \mathcal{H}$, each alcove $A \in \mathcal{A}$ lies in one or the other of the half-spaces defined by H. We say that H **separates** two alcoves A and A' if these alcoves lie in different half-spaces relative to H. For example, H_s separates A_o and sA_o, for each $s \in S_a$.

Note that, for a fixed pair of alcoves, the number of $H \in \mathcal{H}$ which separate them is finite: indeed, a bounded set (such as the line segment joining a pair of points from the two alcoves) obviously meets only finitely many of the parallel hyperplanes $H_{\alpha,k}$ for each fixed α. This allows us to define an integer-valued function on $\widehat{W_a}$ by letting $n(w)$ be the cardinality of the set

$$\mathcal{L}(w) := \{H \in \mathcal{H} \mid H \text{ separates } A_o \text{ and } wA_o\}.$$

In the following section we shall show that the restriction of n to W_a is nothing but the length function ℓ. Of course, $n(1) = 0 = \ell(1)$. It is also easy to see that $n(s) = 1$ if $s \in S_a$, which amounts to showing that $\mathcal{L}(s) = \{H_s\}$: the line segment joining $\lambda \in A_o$ to $s\lambda$ meets no $H \in \mathcal{H}$ other than H_s.

As a further comparison with the length function, observe that $n(w) = n(w^{-1})$: H separates A_o and wA_o if and only if $w^{-1}H$ separates $w^{-1}A_o$ and A_o, so $w^{-1}\mathcal{L}(w) = \mathcal{L}(w^{-1})$.

Note that, if $\mathcal{L}(w)$ is nonempty, then it must contain at least one of the hyperplanes $H_s, s \in S_a$. Otherwise $wA_o \neq A_o$ lies in the same open half-space as A_o relative to each H_s. But A_o is precisely the intersection of these half-spaces, yielding the contradiction $A_o = wA_o$.

As in the case of the length function, it is crucial to determine how the n function changes when we multiply by an element of S_a:

Proposition   *Let $w \in \widehat{W}_a$ and fix $s \in S_a$.*
(a) *H_s belongs to exactly one of the sets $\mathcal{L}(w^{-1}), \mathcal{L}(sw^{-1})$.*
(b) *$s(\mathcal{L}(w^{-1}) \setminus \{H_s\}) = \mathcal{L}(sw^{-1}) \setminus \{H_s\}$.*
(c) *$n(ws) = n(w) - 1$ if $H_s \in \mathcal{L}(w^{-1})$, and $n(ws) = n(w) + 1$ otherwise.*

Proof. (a) Suppose H_s lies in both sets. This implies that wH_s separates A_o from wA_o as well as from wsA_o, so the latter two alcoves lie on the same side of wH_s. This forces A_o and sA_o to lie on the same side of H_s, which is absurd. We get a similar contradiction by supposing that H_s lies in neither set.

(b) Suppose $H = H_{\alpha,k} \neq H_s$ belongs to $\mathcal{L}(w^{-1})$, so $wH \in \mathcal{L}(w)$. Since s fixes H_s, $sH \neq H_s$. We have to show that $sH \in \mathcal{L}(sw^{-1})$. Suppose the contrary: sH does not separate A_o and $sw^{-1}A_o$, hence H does not separate sA_o and $w^{-1}A_o$, hence wH does not separate wsA_o and A_o, i.e., $wH \notin \mathcal{L}(ws)$. But, by assumption, $wH \in \mathcal{L}(w)$, so wH must separate wA_o and wsA_o. Thus $H \in \mathcal{L}(s) = \{H_s\}$ contrary to the choice of H. This proves one inclusion in (b). To get the reverse inclusion, just replace w by ws.

(c) This follows immediately from (a) and (b). □

Corollary   *For any $w \in W_a$, we have $n(w) \leq \ell(w)$.*

Proof. This is clear if $w = 1$. Otherwise let $w = s_1 \cdots s_r$ be a reduced expression, and use induction on r. Part (c) of the proposition shows that the value of n can increase at most 1 each time we multiply by a factor s, so $n(w) \leq r = \ell(w)$. □

4.5   Simple transitivity

In order to prove that $\ell = n$ on W_a, as well as to show that W_a acts simply transitively on $\mathcal{A}$, we want to write down an explicit list of the hyperplanes separating A_o and wA_o. The key step is contained in the following lemma.

Lemma   *If $w \neq 1$ in W_a has a reduced expression $w = s_1 \cdots s_r$, with $s_i \in S_a$, then (setting $H_i := H_{s_i}$) the hyperplanes*

$$H_1, \; s_1 H_2, \; s_1 s_2 H_3, \ldots, \; s_1 \cdots s_{r-1} H_r$$

are all distinct.

Proof. Suppose on the contrary that for some indices $p < q$ we have $s_1 \cdots s_{p-1} H_p = s_1 \cdots s_{q-1} H_q$. Then $H_p = s_p \cdots s_{q-1} H_q$. By Corollary 4.2, this implies $s_p = (s_p \cdots s_{q-1}) s_q (s_{q-1} \cdots s_p)$. Thus $s_p \cdots s_q = s_{p+1} \cdots s_{q-1}$, allowing us to reduce the length of the already reduced expression for w, which is absurd. $\square$

By combining this lemma with Proposition 4.4, we can easily derive the promised conclusions:

Theorem (a) *Let* $w \neq 1$ *in* W_a *have a reduced expression* $w = s_1 \cdots s_r$. *Then we have (setting* $H_i := H_{s_i}$*)*

$$\mathcal{L}(w) = \{H_1, s_1 H_2, s_1 s_2 H_3, \ldots, s_1 \cdots s_{r-1} H_r\}.$$

Moreover, these r *hyperplanes are all distinct.*
(b) *The function* n *on* W_a *coincides with the length function* ℓ.
(c) *The group* W_a *acts simply transitively on* $\mathcal{A}$.

Proof. (a) We have already observed that $\mathcal{L}(s) = \{H_s\}$ when $s \in S_a$. Now proceed by induction on $r = \ell(w)$. When $r > 1$, the induction hypothesis says that

$$\mathcal{L}(s_2 \cdots s_r) = \{H_2, s_2 H_3, \ldots, s_2 \cdots s_{r-1} H_r\}.$$

Moreover, these $r - 1$ hyperplanes are distinct. If H_1 were to occur in this list, then we would have

$$H_1 = s_1 H_1 \in \{s_1 H_2, s_1 s_2 H_3, \ldots, s_1 \cdots s_{r-1} H_r\},$$

contrary to the above lemma. Thus $H_1 \notin \mathcal{L}(s_1 w)$. We now apply Proposition 4.4 (taking $s = s_1$ and replacing w^{-1} there by $s_2 \cdots s_r$). Part (a) forces H_1 to lie in $\mathcal{L}(w)$. Then part (b) forces $\mathcal{L}(w)$ to be the desired set of r hyperplanes (all distinct, by the lemma).
(b) follows immediately from (a).
(c) We already know from 4.3 that W_a acts transitively on $\mathcal{A}$, so it remains to show that no element $w \neq 1$ can fix an alcove, say A_o (without loss of generality). But $w A_o = A_o$ means that $\mathcal{L}(w) = \emptyset$, contrary to (a). $\square$

It is worth observing that the above list of hyperplanes separating A_o from $w A_o$ corresponds precisely to a sequence of r affine reflections whose product is w. This is seen by rewriting w in the form

$$(s_1 \cdots s_{r-1} s_r s_{r-1} \cdots s_1)(s_1 \cdots s_{r-2} s_{r-1} s_{r-2} \cdots s_1) \cdots (s_1 s_2 s_1) s_1.$$

Exercise 1. Show by example that $\widehat{W}_a$ does not in general permute $\mathcal{A}$ simply transitively.

Exercise 2. When restricted to the subgroup W, the length function on W_a agrees with the length function defined in Chapter 1.

Exercise 3. Each $H \in \mathcal{H}$ divides V into 'positive' and 'negative' half-spaces. Associate to any pair of alcoves A, B an integer $d(A, B)$ as follows: To each hyperplane $H \in \mathcal{H}$ which separates A from B, assign $+1$ if B lies on the positive side of H or -1 if B lies on the negative side of H. Sum these values over the (finitely many) hyperplanes separating A from B to get $d(A, B)$. For all $A, B, C \in \mathcal{A}$, prove that

$$d(A, B) + d(B, C) + d(C, A) = 0,$$

by considering each family of parallel hyperplanes separately. If we assign to each alcove A the unique element $w \in W_a$ for which $A = wA_o$, how does $d(A_o, A)$ compare with $\ell(w)$? (See Lusztig [2], 1.4.)

The simple transitivity of W_a on $\mathcal{A}$ has a nice consequence for the structure of $\widehat{W}_a$. If $w \in \widehat{W}_a$, then the transitivity of W_a on alcoves already implies that $wA_o = w'A_o$ for some $w' \in W_a$. Thus $w(w')^{-1}A_o = A_o$. If Ω is the subgroup of $\widehat{W}_a$ stabilizing A_o, this shows that $\widehat{W}_a$ is the product of W_a and Ω. Simple transitivity further implies that $W_a \cap \Omega = 1$, so in fact the product is semidirect and $\Omega \cong \widehat{W}_a/W_a \cong \hat{L}/L$.

4.6   Exchange Condition

Now we are in a position to derive an analogue of the Exchange Condition for finite reflection groups (1.7), equivalent to the Deletion Condition. This in turn will allow us to deduce that (W_a, S_a) is a Coxeter system, just as in 1.9.

Exchange Condition *Let $w \in W_a$ have a reduced expression $w = s_1 \cdots s_r$, with $s_i \in S_a$. If $\ell(ws) < \ell(w)$ ($s \in S_a$), then there exists an index $1 \le i \le r$ for which $ws = s_1 \cdots \hat{s}_i \cdots s_r$.*

Proof. By part (c) of Theorem 4.5,

$$\mathcal{L}(w) = \{H_1, s_1 H_2, \ldots, s_1 \cdots s_{r-1} H_r\}.$$

Therefore

$$\mathcal{L}(w^{-1}) = w^{-1}\mathcal{L}(w) = \{s_r \cdots s_1 H_1, s_r \cdots s_2 H_2, \ldots, s_r H_r\}.$$

Of course, $s_i H_i = H_i$ in each case. By the hypothesis on s together with part (c) of Proposition 4.4, H_s must lie in $\mathcal{L}(w^{-1})$, say $H_s = s_r \cdots s_{i+1} H_i$

for $1 \leq i \leq r$. Then Corollary 4.2 forces $(s_r \cdots s_{i+1})s_i(s_{i+1} \cdots s_r) = s$, or $s_i s_{i+1} \cdots s_r = s_{i+1} \cdots s_r s$. After substitution, this yields $ws = s_1 \cdots \hat{s}_i \cdots s_r$ as desired. □

As discussed in 1.7, this version of the Exchange Condition is equivalent to the Deletion Condition. That condition alone is enough to make the argument in 1.9 work, allowing us to conclude:

Theorem *The pair (W_a, S_a) is a Coxeter system.* □

Exercise. The affine Weyl group of type $\widetilde{A_{n-1}}$ can be realized as a group of permutations of $\mathbf{Z}$, as follows. List the elements of S_a as $s_0, s_1, \ldots, s_{n-1}$, so that, for $i > 0$, s_i is the transposition $(i, i+1)$ in $W = S_n$. Then associate to s_i $(0 \leq i < n)$ the permutation of $\mathbf{Z}$ sending $t \mapsto t - 1$ if $t \equiv i + 1 \pmod{n}$, $t \mapsto t + 1$ if $t \equiv i \pmod{n}$, $t \mapsto t$ otherwise. Prove that this assignment extends to an isomorphism of W_a onto a subgroup of the permutation group, which may be characterized as the set of permutations π satisfying: $\pi(t + n) = \pi(t) + n$ for all $t \in \mathbf{Z}$ and $\sum t = \sum \pi(t)$ (sums taken from 1 to n). How is the length function described in this realization?

4.7 Coxeter graphs and extended Dynkin diagrams

It is not difficult to construct, for each irreducible Weyl group W, the Coxeter graph belonging to the Coxeter group W_a. One just needs to work out the order of $s_\alpha s_{\tilde{\alpha},1}$ for each $\alpha \in \Delta$, to see what new edges and labels occur when the new vertex is adjoined to the Coxeter graph of W. Geometrically, this product of reflections depends only on the angle between the associated hyperplanes, which is the same as the angle between the parallel hyperplanes (through 0) orthogonal to α and $\tilde{\alpha}$. So the calculation is easily done using the data about roots in 2.10. The resulting Coxeter graphs are precisely those occurring in Figure 2 of 2.5, with labels $\widetilde{A_n}$, $\widetilde{B_n}$, etc. Note that $\widetilde{A_1}$ is the only graph involving the label ∞.

In Lie theory it is convenient to codify in an **extended Dynkin diagram** the information about relative root lengths, when an extra root is added to Δ. Since the angles between simple roots are obtuse, it is natural to use $-\tilde{\alpha}$ (often labelled α_0) as the extra root. Indeed, its angle with any simple root is also obtuse: $\tilde{\alpha}$ forms an acute angle with any simple root α, for otherwise $s_\alpha \tilde{\alpha}$ would be a higher root obtained by adding a positive multiple of α. Imitating the construction of Dynkin diagrams (2.9), we obtain the extended diagrams in Figure 1.

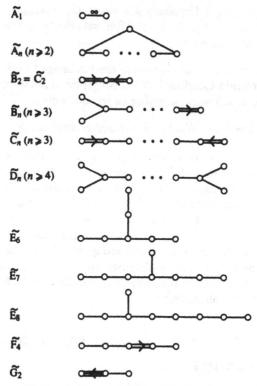

Figure 1: Extended Dynkin diagrams

Example. Consider the crystallographic root system of type B_n described in 2.10. In terms of the standard basis of $\mathbf{R}^n$, the simple roots are

$$\alpha_1 = \varepsilon_1 - \varepsilon_2, \ \alpha_2 = \varepsilon_2 - \varepsilon_3, \ldots, \ \alpha_{n-1} = \varepsilon_{n-1} - \varepsilon_n, \ \alpha_n = \varepsilon_n,$$

while $\tilde{\alpha} = \varepsilon_1 + \varepsilon_2$. Thus for $n = 2$ a double edge joins the vertex associated with the short root α_2 and the new vertex associated with the long root $-\tilde{\alpha}$. When $n \geq 3$ both of these roots are long, so a single edge joins the vertex associated with α_2 and the new vertex.

Exercise. If I is a proper subset of S_a, the resulting 'parabolic subgroup' generated by I is finite. (What can be said about its order?)

4.8   Fundamental domain

Returning to the geometric discussion which led up to the presentation of W_a, we can imitate in a straightforward way the description of fundamental domains for finite reflection groups (1.12). Here the appropriate subset of V is the closure $\overline{A_o}$ of A_o.

Theorem *The closure of A_o is a fundamental domain for the action of W_a on V.*

Proof. Obviously each element of V lies in the closure of at least one alcove. So Proposition 4.3 implies that W_a sends each element of V to some element of $\overline{A_o}$. It just has to be shown that no two distinct elements $\lambda, \mu \in \overline{A_o}$ can be conjugate under W_a. Suppose the contrary: $w\lambda = \mu$ for some $w \in W_a$. We may assume that $\ell(w) > 0$ is as small as possible.

Now we imitate the proof of Theorem 1.12. Find $s \in S_a$ for which $\ell(ws) < \ell(w)$. Thanks to the fact that $\ell(w) = n(w)$ (Theorem 4.5), part (c) of Proposition 4.4 shows that $H_s \in \mathcal{L}(w^{-1})$. Thus H_s separates A_o from $w^{-1}A_o$. Since $\lambda = w^{-1}\mu$ lies in $w^{-1}\overline{A_o}$, we must have $(\lambda, \alpha) \leq 0$ (in case $s = s_\alpha$ for a simple root α) or else $(\lambda, \tilde{\alpha}) \geq 1$ (in case $s = s_{\tilde{\alpha},1}$). But, by assumption, $(\lambda, \alpha) \geq 0$ and $(\lambda, \tilde{\alpha}) \leq 1$, so we get either $(\lambda, \alpha) = 0$ or else $(\lambda, \tilde{\alpha}) = 1$. In either case, $s\lambda = \lambda$, and thus $ws\lambda = \mu$ contrary to the minimality of $\ell(w)$.

(As in the proof of Theorem 1.12, this argument actually shows that the stabilizer of an element of $\overline{A_o}$ is generated by those elements of S_a which it contains.) $\square$

4.9 A formula for the order of W

Comparison of fundamental domains for W_a and for one of the related translation groups leads to a beautiful formula for the order of the Weyl group W, which is independent of the earlier methods developed in 2.11 and 3.9.

Recall the lattices $\hat{L} = \hat{L}(\Phi^\vee)$ and $L = L(\Phi^\vee)$. The index of L in $\hat{L}$ is denoted by f and called the index of connection (2.9). It is easy to compute f in each case from the matrix of Cartan integers (see Humphreys [1], §13); see the table below.

Another ingredient in the formula is the expression of the highest root $\tilde{\alpha}$ as a linear combination of simple roots:

$$\tilde{\alpha} = \sum c_i \alpha_i, \quad \text{where } \Delta = \{\alpha_1, \ldots, \alpha_n\}.$$

The integers c_i are listed in the table below; they can be derived easily from the data in 2.10.

Theorem *If W is an irreducible Weyl group of rank n, then*

$$|W| = n! \, c_1 \cdots c_n \, f,$$

Type	Coefficients of $\tilde{\alpha}$	f
A_n	$1, 1, \ldots, 1$	$n + 1$
B_n	$1, 2, 2, \ldots, 2$	2
C_n	$2, 2, \ldots, 2, 1$	2
D_n	$1, 2, \ldots, 2, 1, 1$	4
E_6	$1, 2, 2, 3, 2, 1$	3
E_7	$2, 2, 3, 4, 3, 2, 1$	2
E_8	$2, 3, 4, 6, 5, 4, 3, 2$	1
F_4	$2, 3, 4, 2$	1
G_2	$3, 2$	1

Table 1: Coefficients of highest root and index of connection

where f is the index of connection and the c_i are the coefficients of the highest root.

Proof. The idea is to compare the volumes of two fundamental domains in V (identified with $\mathbf{R}^n$) relative to the usual Lebesgue measure. Since hyperplanes have measure 0, we do not have to be too precise about the boundaries involved.

Let P be the (open) parallelepiped determined by the basis vectors $\varpi_i^\vee$ of $\hat{L}$ dual to the $\alpha_i \in \Delta$:

$$P := \{\lambda \in V \,|\, 0 < (\lambda, \alpha_i) < 1 \text{ for all } i\}.$$

Since P is bounded by some of the hyperplanes in $\mathcal{H}$, it is clear that P is a union of certain alcoves (including $A_\circ$) and parts of their closures. Moreover, P together with part of its boundary forms a fundamental domain for the translation group corresponding to $\hat{L}$.

We want to compare $A_\circ$ with P. Since $(\varpi_i^\vee, \alpha_j) = \delta_{ij}$, we see that the vertices of $A_\circ$ are the points $(1/c_i)\varpi_i^\vee$ together with 0. Indeed, these are precisely the points obtained by intersecting all but one of the hyperplanes in S_a. An elementary calculation with multiple integrals shows that the volume of the standard n-simplex in $\mathbf{R}^n$ is $1/n!$. If the standard basis is modified by factors $1/c_i$, the resulting volume is just multiplied by these factors. A change of basis to the vectors $\varpi_i^\vee$ in turn modifies the volume by a factor equal to the absolute value of the determinant of the coordinate matrix of these vectors. The same factor changes the volume of the standard unit parallelepiped into the volume of P. So this factor does not affect the ratio:

$$\mathrm{vol}(P)/\mathrm{vol}(A_\circ) = n!\, c_1 \cdots c_n.$$

Another way to compute the ratio of volumes is simply to count how many alcoves are contained in P. We claim that the number of these is $|W|/f$, which will yield the desired formula for $|W|$.

To prove the claim, we first look for all possible elements of $\widehat{W_a}$ which map A_o into P. Any such element is the product of a translation by some $\mu \in \hat{L}$ and an element of W. Let $\lambda \in A_o$ be arbitrary, and $w \in W$. Suppose translation by $\mu = \sum a_i \varpi_i^\vee$ takes $w\lambda$ into P, so $(w\lambda + \mu, \alpha_i)$ lies between 0 and 1 for all i. Now $(w\lambda, \alpha_i) = (\lambda, w^{-1}\alpha_i)$ lies between 0 and 1 (resp. 0 and -1) provided $w^{-1}\alpha_i$ is positive (resp. negative), since $\lambda \in A_o$. Moreover, all such values are attainable as λ varies. On the other hand, $(\mu, \alpha_i) = a_i$ is an integer. Therefore, $(w\lambda + \mu, \alpha_i)$ lies between 0 and 1 precisely when we have $a_i = 0$ (resp. 1) for $w^{-1}\alpha_i > 0$ (resp. < 0). Denote by $\rho_w^\vee$ the sum of all $\varpi_i^\vee$ for which $w^{-1}\alpha_i < 0$. We have shown that $t(\rho_w^\vee)\,w$ is the unique element of $\widehat{W_a}$ involving w which maps A_o into P.

Now for each such element of $\widehat{W_a}$, its product with each of the f elements of Ω, the subgroup stabilizing A_o defined in 4.5, takes A_o to the same alcove in P. On the other hand, two elements of $\widehat{W_a}$ which map A_o to the same alcove obviously differ by an element of Ω. It follows that the number of distinct alcoves in P (those which are images of A_o under W_a) is $|W|/f$, as claimed. $\square$

Remark. The fact that f divides $|W|$ is not *a priori* obvious, but becomes clear when one sees how to locate naturally a subgroup of W isomorphic to Ω. This and other refinements, such as the precise determination of which elements $\rho_w^\vee$ lie in L, may be found in Verma [3] (in a formulation dual to ours) and Iwahori–Matsumoto [1].

Exercise 1. Use the table to check (case-by-case) that the formula in the theorem is in agreement with the earlier formulas for $|W|$ developed in 2.11 and 3.9.

Exercise 2. Verify (using the table) that $f-1$ is the number of coefficients of $\tilde{\alpha}$ equal to 1.

4.10 Groups generated by affine reflections

In this chapter we have constructed a particular class of groups generated by affine reflections, which turn out to be Coxeter groups. It is natural to ask whether some larger class of 'discrete' groups generated by affine reflections may be found. In a word, the answer is no (if the limitation to 'discrete' groups is formulated appropriately). This question is explored exhaustively in Chapters V and VI of Bourbaki [1] (cf. Coxeter [2], Witt [1]). For a very helpful outline (with some proofs filled in) see Brown [1], Chapter VI, §1. Here we offer just a brief guide to what is done in Bourbaki, with references to sections.

(1) Start with an arbitrary collection $\mathcal{H}$ of affine hyperplanes in V, and consider the group G generated by the corresponding reflections.

Using Proposition 4.1, we may enlarge $\mathcal{H}$ if necessary so that G permutes $\mathcal{H}$. To insure that G is not 'too big', we might require that $\mathcal{H}$ be 'locally finite' (any compact set meets only finitely many $H \in \mathcal{H}$). This follows from the formal requirement that G (given the discrete topology) act 'properly' on V: given two compact sets, only finitely many G-translates of the first meet the second. In particular, G will be discrete in the natural topology of Aff(V). [See V, §3.]

(2) With these assumptions in place, the complement of the union of hyperplanes in $\mathcal{H}$ will be open; its connected components may be called 'chambers'. They have naturally defined 'walls' in $\mathcal{H}$. G then permutes the collection of all chambers. (Lower dimensional cells and faces can also be defined in a natural way by systems of equalities and inequalities.) [See V, §1, §2.]

(3) Study of the action of G on chambers leads to a number of familiar-looking conclusions: If C is any fixed chamber, and R the collection of reflections with respect to its walls, then R is finite, G is generated by R and acts simply transitively on the collection of all chambers. Moreover, (G, R) is a Coxeter system, and $\mathcal{H}$ consists of all affine hyperplanes whose reflections lie in G. [See V, §3.]

(4) Note that G might actually be finite (if it fixes a point of V). In general, there are only finitely many parallel classes of hyperplanes in $\mathcal{H}$, and there exist 'special points' where hyperplanes from all these classes intersect. [See V, §3.]

(5) Suppose G is infinite, and 'irreducible' in a natural sense (which does not limit the generality significantly). Then G contains a translation lattice of rank equal to the dimension of V, which is the root lattice of a (crystallographic) root system. Moreover, the stabilizer in G of a special point (which we may as well take to be 0) is a finite subgroup generated by reflections, which normalizes the translation lattice (its root lattice). The conclusion is that G is an affine Weyl group. [See VI, §2.]

Related to this line of reasoning in Bourbaki is a discussion which characterizes such affine reflection groups among all possible Coxeter groups, in terms of the 'geometric representation' and associated bilinear form. [See V, §4 and VI, 4.3.] We shall discuss these matters in Chapters 5 and 6 below.

Notes

For the notion of upper closure in the exercise in 4.3, and various refinements, see Jantzen [3], p. 261.

(4.5) The arrangement of the arguments here was suggested by J.B. Carrell.

(4.6) The exercise is due to Lusztig [3] (see Shi [1], p. 67). He has worked out similar descriptions for other affine Weyl groups; see Bédard [1].

(4.9) This formula is due to Weyl (see Coxeter [1], 11.9). Our treatment follows Verma [3]. The proof in Bourbaki [1], VI, 2.4, prop. 7 (p. 178) uses some facts about Haar measure. For a different proof, based on a comparison of Poincaré series for W and W_a, see Proposition 1.32 of Iwahori–Matsumoto [1].

Part II

General theory of Coxeter groups

Chapter 5

Coxeter groups

Motivated by the examples of finite reflection groups (Chapter 1) and affine Weyl groups (Chapter 4), we embark on the general study of Coxeter groups. After introducing the basic notions in 5.1–5.3, we examine the 'root system' in 5.4–5.7, following Deodhar [4]. This leads to the 'Strong Exchange Condition' (5.8). Then we study the Bruhat ordering in 5.9–5.11. Finally, we look more closely at parabolic subgroups, deriving an inductive formula to express Poincaré series as rational functions in 5.12 and finding a fundamental domain for the action of our group in 5.13.

5.1 Coxeter systems

We define a **Coxeter system** to be a pair (W, S) consisting of a group W and a set of generators $S \subset W$, subject only to relations of the form

$$(ss')^{m(s,s')} = 1,$$

where $m(s, s) = 1$, $m(s, s') = m(s', s) \geq 2$ for $s \neq s'$ in S. In case no relation occurs for a pair s, s', we make the convention that $m(s, s') = \infty$. Formally, W is the quotient F/N, where F is a free group on the set S and N is the normal subgroup generated by all elements

$$(ss')^{m(s,s')}.$$

Call $|S|$ the **rank** of (W, S). The canonical image of S in W is a generating set which might conceivably be smaller than S, but in fact it will soon turn out to be in bijection with S (5.3). In the meantime, we may allow ourselves to write $s \in W$ for the image of $s \in S$, whenever this creates no real ambiguity in the arguments. Moreover, we may refer to W itself as a **Coxeter group**, when the presentation is understood.

105

Although a good part of the theory goes through for arbitrary S, we shall always assume that S is finite.

This definition is of course motivated by the two geometric examples studied earlier: finite groups generated by reflections (Chapter 1) and affine Weyl groups (Chapter 4). However, the subject becomes vastly more general when the choices of the $m(s, s')$ are essentially unrestricted. As a result, the reader may well be skeptical at this point about the depth or interest of such a generalization. It will be seen presently that Coxeter groups do admit a sort of geometric interpretation as groups generated by 'reflections' (in a weak sense), and that they share many interesting features. The special cases just mentioned are the ones most often encountered in applications, but there are further useful classes of Coxeter groups (e.g., the 'hyperbolic' ones, and the 'Weyl groups' associated with Kac–Moody Lie algebras). While the general theory may be regarded at first as mainly a nice unification of existing theories, it also suggests new viewpoints and problems.

To specify a Coxeter system (W, S) is to specify a finite set S and a symmetric matrix M indexed by S, with entries in $\mathbf{Z} \cup \{\infty\}$ subject to the conditions: $m(s, s) = 1$, $m(s, s') \geq 2$ if $s \neq s'$. Equivalently, one can draw an undirected graph Γ with S as vertex set, joining vertices s and s' by an edge labelled $m(s, s')$ whenever this number (∞ allowed) is at least 3. If distinct vertices s and s' are not joined, it is then understood that $m(s, s') = 2$. As a simplifying convention, the label $m(s, s') = 3$ may be omitted. As in 2.1, Γ is called a **Coxeter graph**.

Here are a couple of examples not previously encountered.

Example 1. In case all $m(s, s') = \infty$ when $s \neq s'$, we call W a **universal** Coxeter group (see Dyer [2]). If $|S| = 2$, W is just the infinite dihedral group $\mathcal{D}_\infty$, an affine Weyl group of type $\widetilde{A}_1$.

Example 2. Let $S = \{s_1, s_2, s_3\}$, with $m(s_1, s_2) = 3$, $m(s_1, s_3) = 2$, $m(s_2, s_3) = \infty$, so the Coxeter graph is

$$\circ\!\!-\!\!\circ \overset{\infty}{-\!\!-} \circ$$

The resulting Coxeter group W turns out to be isomorphic to $\mathrm{PGL}(2, \mathbf{Z})$ $= \mathrm{GL}(2, \mathbf{Z})/\{\pm 1\}$. Denote the canonical map $\mathrm{GL}(2, \mathbf{Z}) \to \mathrm{PGL}(2, \mathbf{Z})$ by

$$\begin{pmatrix} a & b \\ c & d \end{pmatrix} \mapsto \begin{bmatrix} a & b \\ c & d \end{bmatrix}.$$

Then send the generators s_1, s_2, s_3 to the respective elements of order 2 in $\mathrm{PGL}(2, \mathbf{Z})$:

$$\begin{bmatrix} 0 & 1 \\ 1 & 0 \end{bmatrix}, \begin{bmatrix} -1 & 1 \\ 0 & 1 \end{bmatrix}, \begin{bmatrix} -1 & 0 \\ 0 & 1 \end{bmatrix}.$$

By checking the orders of the products, we see that this assignment induces a homomorphism $\varphi : W \to \mathrm{PGL}(2, \mathbf{Z})$. The image of φ includes the subgroup $\mathrm{PSL}(2, \mathbf{Z})$ of index 2, since $\varphi(s_1 s_3)$ and $\varphi(s_2 s_3)$ respectively come from elementary matrices

$$\begin{pmatrix} 0 & 1 \\ -1 & 0 \end{pmatrix} \text{ and } \begin{pmatrix} 1 & 1 \\ 0 & 1 \end{pmatrix},$$

which are well known to generate $\mathrm{SL}(2, \mathbf{Z})$. Because $\mathrm{PSL}(2, \mathbf{Z})$ does not contain the images of matrices of determinant -1 representing the s_i, we conclude that φ is surjective. To see that φ is injective, one can use the standard fact that $\mathrm{PSL}(2, \mathbf{Z})$ is the free product of the groups of orders 2 and 3 generated by $\varphi(s_1 s_3)$ and $\varphi(s_1 s_2)$. (W is an example of a 'hyperbolic' Coxeter group; see 6.8 below. It is discussed from several perspectives in Brown [1], pp. 40–46.)

It is notoriously difficult to say much about a group given only by generators and relations — for example, is the group trivial or not? In our case, we can see right away that W has order at least 2. Start with a homomorphism from the free group F onto the multiplicative group $\{1, -1\}$, defined by sending each element of S to -1. It is obvious that all elements $(ss')^{m(s,s')}$ lie in the kernel, so there is an induced epimorphism $\varepsilon : W \to \{1, -1\}$ sending the image of each $s \in S$ to -1. In particular, each of these generators of W does have order 2. The map ε is the generalization for an arbitrary Coxeter group of the sign character of the symmetric group.

Proposition *There is a unique epimorphism $\varepsilon : W \to \{1, -1\}$ sending each generator $s \in S$ to -1. In particular, each s has order 2 in W.* $\square$

Note that when $|S| = 1$, W is just a group of order 2. When $|S| = 2$, W is dihedral, of order $2m(s, s') \leq \infty$ if $S = \{s, s'\}$. So we are already well acquainted with these types of Coxeter groups in the guise of reflection groups.

Exercise 1. Denote the kernel of ε by W^+. If $S = \{s_1, \ldots, s_n\}$, prove that W^+ is generated by the elements $s_i s_n$ ($1 \leq i \leq n - 1$).

Exercise 2. If W has rank n and all $m(s, s')$, $s \neq s'$, are even, then $|W| \geq 2^n$.

5.2 Length function

Since the generators $s \in S$ have order 2 in W, each $w \neq 1$ in W can be written in the form $w = s_1 s_2 \cdots s_r$ for some s_i (not necessarily distinct) in S. If r is as small as possible, call it the **length** of w, written $\ell(w)$, and call any expression of w as a product of r elements of S a **reduced expression**. By convention, $\ell(1) = 0$. More formally, a reduced expression should be viewed as an ordered r-tuple $(s_1, \ldots, s_r)$. Note that the lengths of partial products are predictable when $w = s_1 \cdots s_r$ is reduced: $\ell(s_1 \cdots s_{r-1}) = r - 1$, $\ell(s_2 \cdots s_{r-1}) = r - 2$, etc. However, the length function has its subtleties, because a typical element of W may have numerous reduced expressions.

Exercise. Prove that W is of 'universal' type (5.1) if and only if each element has a unique reduced expression.

Here are some elementary properties of the length function:

(L1) $\ell(w) = \ell(w^{-1})$. [If $w = s_1 \cdots s_r$, $w^{-1} = s_r \cdots s_1$, so $\ell(w^{-1}) \leq \ell(w)$, and similarly for w^{-1} in place of w.]

(L2) $\ell(w) = 1$ if and only if $w \in S$.

(L3) $\ell(ww') \leq \ell(w) + \ell(w')$. [If $w = s_1 \cdots s_p$ and $w' = s_1' \cdots s_q'$, then the product $ww' = s_1 \cdots s_p s_1' \cdots s_q'$ has length at most $p + q$.]

(L4) $\ell(ww') \geq \ell(w) - \ell(w')$. [Apply (L3) to the pair $ww', (w')^{-1}$, then use (L1).]

(L5) $\ell(w) - 1 \leq \ell(ws) \leq \ell(w) + 1$, for $s \in S$ and $w \in W$. [Use (L3) and (L4).]

Proposition   *The homomorphism $\varepsilon : W \to \{1, -1\}$ of 5.1 is given by $\varepsilon(w) = (-1)^{\ell(w)}$. As a result, $\ell(ws) = \ell(w) \pm 1$, for all $s \in S, w \in W$, and similarly for $\ell(sw)$.*

Proof. Write a reduced expression $w = s_1 \cdots s_r$. Then

$$\varepsilon(w) = \varepsilon(s_1) \cdots \varepsilon(s_r) = (-1)^r = (-1)^{\ell(w)},$$

as required. Now $\varepsilon(ws) = -\varepsilon(w)$ implies that $\ell(ws) \neq \ell(w)$. By property (L5) above, the lengths must differ by precisely 1. $\square$

In our study of Coxeter groups (as in the special cases treated earlier), we shall often prove theorems by induction on $\ell(w)$. It will therefore be essential to understand the precise relationship between $\ell(w)$ and $\ell(ws)$ (or $\ell(sw)$). For this we need a way to represent W concretely.

5.3   Geometric representation of W

Given a Coxeter system (W, S), it is too much to expect a faithful representation of W as a group generated by (orthogonal) reflections in a euclidean space. But we can get a reasonable substitute if we redefine a **reflection** to be merely a linear transformation which fixes a hyperplane pointwise and sends some nonzero vector to its negative. The idea is to begin with a vector space V over $\mathbf{R}$, having a basis $\{\alpha_s | s \in S\}$ in one-to-one correspondence with S, and then to impose a geometry on V in such a way that the 'angle' between α_s and $\alpha_{s'}$ will be compatible with the given $m(s, s')$, i.e., with the previously studied geometry of dihedral groups. Accordingly, we define a symmetric bilinear form B on V by requiring:

$$B(\alpha_s, \alpha_{s'}) = -\cos \frac{\pi}{m(s, s')}.$$

(This expression is interpreted to be -1 in case $m(s, s') = \infty$.) Evidently $B(\alpha_s, \alpha_s) = 1$, while $B(\alpha_s, \alpha_{s'}) \leq 0$ if $s \neq s'$. Since α_s is non-isotropic, the subspace H_s orthogonal to α_s relative to B is complementary to the line $\mathbf{R}\alpha_s$.

For each $s \in S$ we can now define a reflection $\sigma_s : V \to V$ by the rule:

$$\sigma_s \lambda = \lambda - 2B(\alpha_s, \lambda)\alpha_s.$$

Clearly $\sigma_s \alpha_s = -\alpha_s$, while q_s fixes H_s pointwise. In particular, we see that σ_s has order 2 in $\mathrm{GL}(V)$.

A quick calculation (left to the reader) shows that σ_s preserves the form B, i.e., $B(\sigma_s \lambda, \sigma_s \mu) = B(\lambda, \mu)$ for all $\lambda, \mu \in V$. As a result, each element of the subgroup of $\mathrm{GL}(V)$ generated by the $\sigma_s (s \in S)$ will also preserve B.

Our first task is to show that there exists a homomorphism from W onto this linear group, sending s to σ_s. For this it is enough to check that

$$(\sigma_s \sigma_{s'})^{m(s,s')} = 1 \text{ whenever } s \neq s'.$$

Set $m := m(s, s')$ and consider first the two-dimensional subspace $V_{s,s'} := \mathbf{R}\alpha_s \oplus \mathbf{R}\alpha_{s'}$. We claim that *the restriction of B to $V_{s,s'}$ is positive semidefinite, and moreover is nondegenerate precisely when $m < \infty$.* To check the first part, just take any $\lambda = a\alpha_s + b\alpha_{s'}$ ($a, b \in \mathbf{R}$), and compute

$$B(\lambda, \lambda) = a^2 - 2ab\cos(\pi/m) + b^2 = (a - b\cos(\pi/m))^2 + b^2 \sin^2(\pi/m) \geq 0.$$

In turn, the form is positive definite on $V_{s,s'}$ if $\sin(\pi/m) \neq 0$, i.e., $m < \infty$ (whereas otherwise the nonzero vector $\alpha_s + \alpha_{s'}$ is isotropic).

Having seen precisely how the form B behaves on $V_{s,s'}$, we note further that σ_s and $\sigma_{s'}$ leave $V_{s,s'}$ stable: just look at the defining formula for each reflection. So it makes sense to calculate the order of $\sigma_s \sigma_{s'}$ viewed as an operator on $V_{s,s'}$. Two cases are possible:

(a) $m < \infty$. Here the form is positive definite, so we find ourselves in the familiar situation of the euclidean plane. Both σ_s and $\sigma_{s'}$ act as orthogonal reflections. Since $B(\alpha_s, \alpha_{s'}) = -\cos(\pi/m) = \cos(\pi - (\pi/m))$, the angle between the rays $\mathbf{R}^+\alpha_s$ and $\mathbf{R}^+\alpha_{s'}$ is $\pi - (\pi/m)$, forcing the angle between the two reflecting lines to be π/m. From our previous study of dihedral groups (1.1), we recognize $\sigma_s\sigma_{s'}$ as a rotation through the angle $2\pi/m$; it therefore has order m.

(b) $m = \infty$. Here $B(\alpha_s, \alpha_{s'}) = -1$. If $\lambda = \alpha_s + \alpha_{s'}$, $B(\lambda, \alpha_s) = 0 = B(\lambda, \alpha_{s'})$, so that both σ_s and $\sigma_{s'}$ fix λ. In turn, $\sigma_s\sigma_{s'}\alpha_s = \sigma_s(\alpha_s + 2\alpha_{s'}) = 3\alpha_s + 2\alpha_{s'} = 2\lambda + \alpha_s$, and by iteration, $(\sigma_s\sigma_{s'})^k\alpha_s = 2k\lambda + \alpha_s$ $(k \in \mathbf{Z})$. This implies that $\sigma_s\sigma_{s'}$ has infinite order on $V_{s,s'}$ (and therefore also on V).

In case (a), the fact that B is nondegenerate on $V_{s,s'}$ implies that V is the orthogonal direct sum of $V_{s,s'}$ and its orthogonal complement; evidently both σ_s and $\sigma_{s'}$ fix the latter subspace pointwise. Thus $\sigma_s\sigma_{s'}$ also has order m on V. To summarize:

Proposition   *There is a unique homomorphism $\sigma : W \to \mathrm{GL}(V)$ sending s to σ_s, and the group $\sigma(W)$ preserves the form B on V. Moreover, for each pair $s, s' \in S$, the order of ss' in W is precisely $m(s, s')$.*
□

This last observation removes any possible ambiguity in the status of the generators $s \in S$: if $s \neq s'$ in the subset S of the free group F, then also $s \neq s'$ in W, as promised in 5.1, and the subgroup of W generated by s, s' is dihedral of order $2m(s, s')$. Now we know that W is not 'too small'. It remains to be seen that W is not 'too big', i.e., that σ has trivial kernel (Corollary 5.4 below). This will require a closer study of the action on V.

For convenience we shall refer to the homomorphism σ as the **geometric representation** of W. (However, it should be emphasized that there may be other interesting ways to represent W as a group generated by 'reflections', e.g., acting in a hyperbolic space. See Vinberg [1]–[5].)

Question. If W is an affine Weyl group, how does the geometric representation compare with the action on euclidean space described in Chapter 4? (This will be discussed in 6.5.)

Exercise. Prove that $s, s' \in S$ are conjugate in W if and only if the following condition is satisfied: (*) There are elements $s = s_1, s_2, \ldots, s_k = s'$ in S for which every $s_i s_{i+1}$ has (finite) odd order.

($\Leftarrow$) In case $w = ss'$ itself has odd order $2p + 1$, note that $w^p s w^{-p} = s'$. Iterate!

($\Rightarrow$) Fix $s \in S$, and consider the set S' of all s' satisfying (*). It must be shown that no element of $S'' := S \setminus S'$ is conjugate to s. Define $f : S \to \{1, -1\}$ by $f(S') = 1$, $f(S'') = -1$. Show that f induces a homomorphism from W to $\{1, -1\}$. Then all conjugates of s must lie in $\mathrm{Ker}\, f$.

5.4 Positive and negative roots

In this section we obtain a precise criterion for $\ell(ws)$ to be greater or smaller than $\ell(w)$, in terms of the action of W on V. This will be the key to all further combinatorial properties of W relative to the generating set S. To avoid cumbersome notation, we may write $w(\alpha_s)$ in place of $\sigma(w)(\alpha_s)$.

First we introduce the **root system** Φ of W, consisting of a set of unit vectors in V permuted by W. Define Φ to be the collection of all vectors $w(\alpha_s)$, where $w \in W$ and $s \in S$. These are unit vectors, because W preserves the form B on V. Note that $\Phi = -\Phi$, since $s(\alpha_s) = -\alpha_s$. If α is any root, we can write it uniquely in the form

$$\alpha = \sum_{s \in S} c_s \alpha_s \ (c_s \in \mathbf{R}).$$

Call α **positive** (resp. **negative**) and write $\alpha > 0$ (resp. $\alpha < 0$) if all $c_s \geq 0$ (resp. all $c_s \leq 0$). For example, each α_s is positive. Write Φ^+ and Φ^- for the respective sets of positive and negative roots. It will be an immediate consequence of the theorem below that these sets exhaust Φ.

Note that, in contrast to the situation in Chapter 1, we have in effect specified once and for all a set of 'simple' roots.

We also have to introduce at this point the **parabolic subgroup** W_I of W, defined as in 1.10 to be the subgroup generated by a given subset $I \subset S$. (More generally, we refer to any conjugate of such a subgroup as a parabolic subgroup.) In the following section, W_I will be seen to be a Coxeter group in its own right. For the present, we just note that it has a length function ℓ_I relative to the generating set of involutions I. It is clear that $\ell(w) \leq \ell_I(w)$ for all $w \in W_I$. (It will be seen in 5.5 that equality holds.)

Theorem   Let $w \in W$ and $s \in S$. If $\ell(ws) > \ell(w)$, then $w(\alpha_s) > 0$. If $\ell(ws) < \ell(w)$, then $w(\alpha_s) < 0$.

Proof. Observe that the second statement follows from the first, applied to ws in place of w: indeed, if $\ell(ws) < \ell(w)$, then $\ell((ws)s) > \ell(ws)$, forcing $ws(\alpha_s) > 0$, i.e., $w(-\alpha_s) > 0$, or $w(\alpha_s) < 0$.

To prove the first statement, we proceed by induction on $\ell(w)$. In case $\ell(w) = 0$, we have $w = 1$, and there is nothing to prove. If $\ell(w) > 0$, we can find an $s' \in S$ for which $\ell(ws') = \ell(w) - 1$, say by choosing s' to be the last factor in a reduced expression for w. Since $\ell(ws) > \ell(w)$ by assumption, we see that $s \neq s'$. Set $I := \{s, s'\}$, so that W_I is dihedral. Now we make a crucial choice within the coset wW_I. Consider the set

$$A := \{v \in W \mid v^{-1}w \in W_I \text{ and } \ell(v) + \ell_I(v^{-1}w) = \ell(w)\}.$$

Evidently $w \in A$. Choose $v \in A$ for which $\ell(v)$ is as small as possible, and write $v_I := v^{-1}w \in W_I$. Thus $w = vv_I$, with $\ell(w) = \ell(v) + \ell_I(v_I)$. The strategy now is to analyze how each of v and v_I acts on roots.

Observe that $ws' \in A$: Indeed, $(s'w^{-1})w = s'$ lies in W_I, while $\ell(ws') + \ell_I(s') = (\ell(w) - 1) + 1 = \ell(w)$. The choice of v therefore forces $\ell(v) \le \ell(ws') = \ell(w) - 1$. This will allow us to apply the induction hypothesis to the pair v, s. But for this we need to compare the lengths of v and vs.

Suppose it were true that $\ell(vs) < \ell(v)$, i.e., $\ell(vs) = \ell(v) - 1$. Then we could calculate as follows:

$$
\begin{aligned}
\ell(w) &\le \ell(vs) + \ell((sv^{-1})w) \quad &&[\text{use (L3) from 5.2}] \\
&\le \ell(vs) + \ell_I(sv^{-1}w) \quad &&[\text{since } sv^{-1}w \in W_I \text{ and } \ell \le \ell_I] \\
&= (\ell(v) - 1) + \ell_I(sv^{-1}w) \\
&\le \ell(v) - 1 + \ell_I(v^{-1}w) + 1 \\
&= \ell(v) + \ell_I(v^{-1}w) \\
&= \ell(w).
\end{aligned}
$$

So equality holds throughout, forcing $\ell(w) = \ell(vs) + \ell_I((sv^{-1})w)$ and therefore $vs \in A$, contrary to $\ell(vs) < \ell(v)$. This contradiction shows that we must instead have $\ell(vs) > \ell(v)$. By induction, we obtain: $v(\alpha_s) > 0$. An entirely similar argument shows that $\ell(vs') > \ell(v)$, whence $v(\alpha_{s'}) > 0$.

Since $w = vv_I$, we will be done if we can show that v_I maps α_s to a nonnegative linear combination of α_s and $\alpha_{s'}$.

We claim that $\ell_I(v_I s) \ge \ell_I(v_I)$. Otherwise we would have:

$$
\ell(ws) = \ell(vv^{-1}ws) \le \ell(v) + \ell(v^{-1}ws) = \ell(v) + \ell(v_I s)
$$

$$
\le \ell(v) + \ell_I(v_I s) < \ell(v) + \ell_I(v_I) = \ell(w),
$$

contrary to $\ell(ws) > \ell(w)$. In turn, it follows that any reduced expression for v_I in W_I (an alternating product of factors s and s') must end in s'. Consider the two possible cases:

(a) If $m(s, s') = \infty$, an easy direct calculation shows that $v_I(\alpha_s) = a\alpha_s + b\alpha_{s'}$, with $a, b \ge 0$ and $|a - b| = 1$. Indeed, $B(\alpha_s, \alpha_{s'}) = -1$, so that $s'(\alpha_s) = \alpha_s + 2\alpha_{s'}, ss'(\alpha_s) = 2\alpha_{s'} + 3\alpha_s, s'ss'(\alpha_s) = 3\alpha_s + 4\alpha_{s'}$, and so on.

(b) If $m := m(s, s') < \infty$, notice that $\ell_I(v_I) < m$. Indeed, m is clearly the maximum possible value of ℓ_I, and an element of length m in W_I has a reduced expression ending with s. So v_I can be written as a product of fewer than $m/2$ terms ss', possibly preceded by one factor s'. Direct calculation will now show that $v_I(\alpha_s)$ is a nonnegative linear combination of α_s and $\alpha_{s'}$. (A rough sketch should make the argument transparent.) Recall that we are now working in the euclidean plane, with unit vectors α_s and $\alpha_{s'}$ at an angle of $\pi - \pi/m$, and ss' rotates

α_s through an angle of $2\pi/m$ toward $\alpha_{s'}$. So the rotations involved in v_I move α_s through at most an angle of $\pi - 2\pi/m$, still within the positive cone defined by α_s and $\alpha_{s'}$. If v_I further involves a reflection corresponding to s', the resulting vector still lies within this positive cone, because the angle between α_s and the reflecting line is $(\pi/2) - (\pi/m)$. $\square$

Corollary *The representation $\sigma : W \to GL(V)$ is faithful.*

Proof. Let $w \in \text{Ker } \sigma$. If $w \neq 1$, there exists $s \in S$ for which $\ell(ws) < \ell(w)$. The theorem says that $w(\alpha_s) < 0$. But $w(\alpha_s) = \alpha_s > 0$, which is a contradiction. $\square$

5.5 Parabolic subgroups

With Theorem 5.4 in hand, we can get more precise information about the internal structure of W. First we want to clarify (as promised) the nature of the parabolic subgroups W_I $(I \subset S)$. The set I and the corresponding values $m(s, s')$ give rise to an abstractly defined Coxeter group $\overline{W_I}$, to which our previous results apply. In particular, $\overline{W_I}$ has a geometric representation of its own. This can obviously be identified with the action of the group generated by all σ_s $(s \in I)$ on the subspace V_I of V spanned by all α_s $(s \in I)$, since the bilinear form B restricted to V_I agrees with the form B_I defined by $\overline{W_I}$. The group generated by these σ_s is just the restriction to V_I of the group $\sigma(W_I)$. On the other hand, $\overline{W_I}$ maps canonically onto W_I, yielding a commutative triangle:

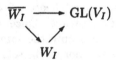

Since the map $\overline{W_I} \to GL(V_I)$ is injective by 5.4, we conclude that W_I is isomorphic to $\overline{W_I}$ and is therefore itself a Coxeter group.

Theorem (a) *For each subset I of S, the pair (W_I, I) with the given values $m(s, s')$ is a Coxeter system.*

(b) *Let $I \subset S$. If $w = s_1 \cdots s_r$ $(s_i \in S)$ is a reduced expression, and $w \in W_I$, then all $s_i \in I$. In particular, the function ℓ agrees with ℓ_I on W_I, and $W_I \cap S = I$.*

(c) *The assignment $I \mapsto W_I$ defines a lattice isomorphism between the collection of subsets of S and the collection of subgroups W_I of W.*

(d) *S is a minimal generating set for W.*

Proof. We have just verified (a). For (b), use induction on $\ell(w)$, noting that $\ell(1) = 0 = \ell_I(1)$. Suppose $w \neq 1$, and set $s = s_r$. According to

Theorem 5.4, $w(\alpha_s) < 0$. Since $w \in W_I$, we can also write $w = t_1 \cdots t_q$ with all $t_i \in I$. Therefore

$$w(\alpha_s) = \alpha_s + \sum_{i=1}^{q} c_i \alpha_{t_i} \ (c_i \in \mathbf{R}).$$

Because $w(\alpha_s) < 0$, we must have $s = t_i$ for some i, forcing $s \in I$. In turn, $ws = s_1 \cdots s_{r-1} \in W_I$, and the expression is reduced. By induction, all $s_i \in I$. The remaining assertions of (b) are clear.

To prove (c), suppose $I, J \subset S$. If $W_I \subset W_J$, then $I = W_I \cap S \subset W_J \cap S = J$, thanks to (b). Thus $I \subset J$ (resp. $I = J$) if and only if $W_I \subset W_J$ (resp. $W_I = W_J$). It is clear that $W_{I \cup J}$ is the subgroup of W generated by W_I and W_J. On the other hand, (b) implies that $W_{I \cap J} = W_I \cap W_J$. This yields the desired lattice isomorphism. To prove (d), suppose that a subset I of S generates W, so $W_I = W = W_S$. According to (c), $I = S$. □

Example. When the Coxeter group in question is an affine Weyl group W_a associated with a Weyl group W (Chapter 4), W itself is a parabolic subgroup of W_a: its Coxeter graph is obtained from that of W_a by removing a single vertex. In particular, the length functions of these groups are compatible.

5.6   Geometric interpretation of the length function

Our next goal is to extract from Theorem 5.4 a more precise description of the way in which W permutes Φ. Once we have this information in hand, we can explore more deeply the internal structure of W itself. Recall that Φ is the disjoint union of the sets Φ^+ and Φ^- of positive and negative roots. For brevity, write $\Pi = \Phi^+$.

Proposition (a) *If $s \in S$, then s sends α_s to its negative, but permutes the remaining positive roots.*

(b) *For any $w \in W$, $\ell(w)$ equals the number of positive roots sent by w to negative roots.*

Proof. Note that part (a) is a special case of part (b); but it is needed in the proof of (b).

(a) Suppose $\alpha > 0$, but $\alpha \neq \alpha_s$. Since all roots are unit vectors, α cannot be a multiple of α_s. We can therefore write

$$\alpha = \sum_{t \in S} c_t \alpha_t,$$

where all coefficients are nonnegative and some $c_t > 0$, $t \neq s$. Applying s to α only modifies this sum by adding some constant multiple of α_s, so the coefficient of α_t remains strictly positive. It follows that $s(\alpha)$ cannot be a negative root, so it lies in Π and is obviously distinct from α_s. Thus $s(\Pi \setminus \{\alpha_s\}) \subset \Pi \setminus \{\alpha_s\}$. Apply s to both sides to get the reverse inclusion.

(b) If $w \in W$, define $n(w)$ to be the number of positive roots sent by w to negative roots, so

$$n(w) = \operatorname{Card} \Pi(w), \quad \text{where } \Pi(w) := \Pi \cap w^{-1}(-\Pi).$$

(It is not instantly obvious that $n(w)$ is finite, but this will follow from the proof that $n(w) = \ell(w)$.) Notice that part (a) implies that $n(s) = 1$ for $s \in S$.

To see that $n(w)$ behaves like the length function, we first verify that, for $s \in S, w \in W$, the condition $w(\alpha_s) > 0$ implies $n(ws) = n(w) + 1$, whereas $w(\alpha_s) < 0$ implies $n(ws) = n(w) - 1$. Indeed, if $w(\alpha_s) > 0$, part (a) implies that $\Pi(ws)$ is the disjoint union of $s(\Pi(w))$ and $\{\alpha_s\}$. Similarly, if $w(\alpha_s) < 0$, we get $\Pi(ws) = s(\Pi(w) \setminus \{\alpha_s\})$, with $\alpha_s \in \Pi(w)$.

Now we proceed by induction on $\ell(w)$ to prove that $n(w) = \ell(w)$ for all $w \in W$. This is clear if $\ell(w) = 0$, and also (by part (a)) if $\ell(w) = 1$. Theorem 5.4 says that $\ell(ws) = \ell(w) + 1$ (resp. $\ell(w) - 1$) just when $w(\alpha_s) > 0$ (resp. < 0). Combining this with the preceding paragraph and the induction hypothesis completes the proof. $\square$

As in the case of finite reflection groups, part (a) of the proposition is invoked frequently, usually as a device for recognizing that a positive root obtained in the course of an argument is none other than α_s (because s sends it to a negative root).

Exercise 1. Given a reduced expression $w = s_1 \cdots s_r$ ($s_i \in S$), set $\alpha_i := \alpha_{s_i}$ and $\beta_i := s_r s_{r-1} \cdots s_{i+1}(\alpha_i)$, interpreting β_r to be α_r. Prove that $\Pi(w)$ consists of the r distinct positive roots $\beta_1, \ldots, \beta_r$.

Exercise 2. If W is infinite, prove that the length function takes arbitrarily large values, hence that Φ is infinite. (Therefore the scalar $-1 \in GL(V)$ does not lie in $\sigma(W)$.) If W is finite, prove that there is one and only one element $w_o \in W$ of maximum length, and that w_o maps Π onto $-\Pi$.

Exercise 3. Use the fact that $\ell(w) = n(w)$ to give another proof of part (b) of Theorem 5.5. [Note that for $w \in W_I$, $n(w) \geq n_I(w)$ is clear, if n_I has the obvious meaning.]

5.7 Roots and reflections

By the way $\sigma : W \to \mathrm{GL}(V)$ was defined, each $s \in S$ acts on V as a reflection. More generally, we can associate a reflection in $\mathrm{GL}(V)$ with each root $\alpha \in \Phi$, as follows. Say $\alpha = w(\alpha_s)$ for some $w \in W, s \in S$. Consider how wsw^{-1} acts on V:

$$
\begin{aligned}
wsw^{-1}(\lambda) &= w[w^{-1}(\lambda) - 2B(w^{-1}(\lambda), \alpha_s)\alpha_s] \\
&= \lambda - 2B(w^{-1}(\lambda), \alpha_s)\, w(\alpha_s) \\
&= \lambda - 2B(\lambda, w(\alpha_s))\, w(\alpha_s) \\
&= \lambda - 2B(\lambda, \alpha)\alpha.
\end{aligned}
$$

It follows that wsw^{-1} depends only on α, not on the choice of w and s. So we may denote it by s_α. Moreover, s_α acts on V as a reflection, sending α to $-\alpha$ and fixing pointwise the hyperplane orthogonal to α. Of course, both α and $-\alpha$ determine the same reflection $s_\alpha = s_{-\alpha}$. Denote by T the set of all reflections s_α, $\alpha \in \Phi$. Thus

$$
T = \bigcup_{w \in W} wSw^{-1}.
$$

In order to pass back and forth freely between roots and reflections, we should observe that the correspondence $\alpha \mapsto s_\alpha$ is bijective (for $\alpha \in \Pi$). Indeed, suppose that $s_\alpha = s_\beta$. From the above formula for s_α (with $\lambda = \beta$) we get $\beta = B(\beta, \alpha)\alpha$, forcing $\alpha = \beta$ since both are unit vectors in Π.

One other observation is useful:

Lemma If $\alpha, \beta \in \Phi$ and $\beta = w(\alpha)$ for some $w \in W$, then $ws_\alpha w^{-1} = s_\beta$.

Proof. This is immediate from the above formula for a reflection and the fact that B is W-invariant. $\square$

The following proposition generalizes Theorem 5.4 to arbitrary reflections.

Proposition Let $w \in W, \alpha \in \Pi$. Then $\ell(ws_\alpha) > \ell(w)$ if and only if $w(\alpha) > 0$.

Proof. As in the proof of Theorem 5.4, it will be enough to verify the 'only if' part. Proceed by induction on $\ell(w)$, the case $\ell(w) = 0$ being trivial. If $\ell(w) > 0$, there exists $s \in S$ such that $\ell(sw) < \ell(w)$. Then $\ell((sw)s_\alpha) = \ell(s(ws_\alpha)) \geq \ell(ws_\alpha) - 1 > \ell(w) - 1 = \ell(sw)$. By induction, $sw(\alpha) > 0$. Suppose $w(\alpha) < 0$. The only negative root made positive by s is $-\alpha_s$ (5.6), so $w(\alpha) = -\alpha_s$. But then $sw(\alpha) = \alpha_s$ would imply $(sw)s_\alpha(sw)^{-1} = s$ (by the above lemma), whence $ws_\alpha = sw$. This contradicts $\ell(ws_\alpha) > \ell(w) > \ell(sw)$. As a result, $w(\alpha)$ must be positive. $\square$

5.8 Strong Exchange Condition

We are now able to prove a key fact about the nature of reduced expressions in W, which is at the heart of what it means to be a Coxeter group.

Theorem *Let $w = s_1 \cdots s_r$ ($s_i \in S$), not necessarily a reduced expression. Suppose a reflection $t \in T$ satisfies $\ell(wt) < \ell(w)$. Then there is an index i for which $wt = s_1 \cdots \widehat{s_i} \cdots s_r$ (omitting s_i). If the expression for w is reduced, then i is unique.*

Proof. Write $t = s_\alpha$ (say $\alpha > 0$). Since $\ell(wt) < \ell(w)$, Proposition 5.7 forces $w(\alpha) < 0$. Because $\alpha > 0$, there exists an index $i \le r$ for which $s_{i+1} \cdots s_r(\alpha) > 0$ but $s_i s_{i+1} \cdots s_r(\alpha) < 0$. According to part (a) of Proposition 5.6, the only positive root which s_i sends to a negative root is α_{s_i}, so $s_{i+1} \cdots s_r(\alpha) = \alpha_{s_i}$. Now Lemma 5.7 implies $(s_{i+1} \cdots s_r)t(s_r \cdots s_{i+1}) = s_i$, or $wt = s_1 \cdots \widehat{s_i} \cdots s_r$ as required.

In case $\ell(w) = r$, consider what would happen if there were distinct indices $i < j$ such that $wt = s_1 \cdots \widehat{s_i} \cdots s_j \cdots s_r = s_1 \cdots s_i \cdots \widehat{s_j} \cdots s_r$. After cancelling, this gives $s_{i+1} \cdots s_j = s_i \cdots s_{j-1}$, or $s_i \cdots s_j = s_{i+1} \cdots s_{j-1}$, allowing us to write $w = s_1 \cdots \widehat{s_i} \cdots \widehat{s_j} \cdots s_r$. This contradicts the assumption that $\ell(w) = r$. □

Exercise 1. Prove a version of the theorem in which the hypothesis reads: $\ell(tw) < \ell(w)$.

We shall refer to the main assertion of the theorem as the **Strong Exchange Condition**. If t is required to lie in S, the resulting weaker statement is called the **Exchange Condition**, generalizing what we proved in the case of finite reflection groups (1.7) and affine Weyl groups (4.6):

Corollary (a) *Suppose $w = s_1 \cdots s_r$ ($s_i \in S$), with $\ell(w) < r$. Then there exist indices $i < j$ for which $w = s_1 \cdots \widehat{s_i} \cdots \widehat{s_j} \cdots s_r$. (This is called the **Deletion Condition**.)*

(b) *If $w = s_1 \cdots s_r$ ($s_i \in S$), then a reduced expression for w may be obtained by omitting certain s_i (an even number, in fact).*

Proof. (a) The hypothesis implies that there exists an index j for which $\ell(w's_j) < \ell(w')$, where $w' := s_1 \cdots s_{j-1}$. Applying the Exchange Condition to the pair w', s_j, we get $w's_j = s_1 \cdots \widehat{s_i} \cdots s_{j-1}$, or $w = s_1 \cdots \widehat{s_i} \cdots \widehat{s_j} \cdots s_r$.

(b) This follows inductively from (a). □

This brings us full circle: recall that the proof in 1.9 shows that any group generated by a set S of involutions and satisfying the Deletion Condition must be a Coxeter group. The theory developed so far in this chapter should, in principle, allow us to answer any reasonable question about Coxeter groups. In practice, some ingenuity is often required. For example, it turns out to be true that the subset of S involved in writing a reduced expression for an element $w \in W$ is independent of the particular reduced expression chosen. A related fact is the equality $W_I \cap W_J = W_{I \cap J}$. The reader might think about how to prove these using the Exchange Condition (see 5.10 below for a less direct approach).

Exercise 2. Let $I \subset S$. Prove that W_I is normal in W if and only if all $s \in S \setminus I$ commute with all $s' \in I$. In terms of the Coxeter graph, this means that I corresponds to a union of some connected components. [Use the Exchange Condition to analyze the length of $ss's$ in W_I.]

Exercise 3. Suppose $w \in W$ acts on V as a reflection, in the sense that there exists a unit vector $\alpha \in V$ for which $w(\lambda) = \lambda - 2B(\lambda, \alpha)\alpha$ for all $\lambda \in V$. Prove that α is a root and $w = s_\alpha$. [First show that, if $s \in S$ and $\ell(ws) < \ell(w)$, then either $\ell(sws) = \ell(w) - 2$ or else $w(\alpha_s) = -\alpha_s$, using just the fact that $w^2 = 1$: find a reduced expression $w = s_1 \cdots s_r$ with $s_r = s$, so $w = s_r \cdots s_1$ is also reduced, and use the Exchange Condition together with 5.6. Now proceed by induction on $\ell(w)$, to show that $w(\beta) = -\beta$ for some root β, whence $\beta = \alpha$ or $-\alpha$, and w is the reflection belonging to α.]

Exercise 4. If $I \subset S$, set $T_I := \bigcup_{w \in W_I} wIw^{-1}$ (the set of reflections in the Coxeter group W_I). Prove that $T \cap W_I = T_I$. [If $t \in T \cap W_I$, write $t = wsw^{-1} = s_1 \cdots s_r$, with $s \in S$, $w \in W$, $s_i \in I$ for all i, and $\ell(ws) > \ell(w)$. Use the Exchange Condition to show that $t = (w')^{-1}s'w'$ for some $s' = s_i, w' = s_{i+1} \cdots s_r$.]

5.9   Bruhat ordering

Among the possible ways to partially order W in a way compatible with the length function, the most useful has proven to be the Bruhat ordering, defined as follows.

As before, T is the set of reflections in W with respect to roots. Write $w' \to w$ if $w = w't$ for some $t \in T$ with $\ell(w) > \ell(w')$. Then define $w' < w$ if there is a sequence $w' = w_0 \to w_1 \to \ldots \to w_m = w$. It is clear that the resulting relation $w' \leq w$ is a partial ordering of W (reflexive, antisymmetric, transitive), with 1 as the unique minimal element. Following Verma [2], we call it the **Bruhat ordering**. The terminology is motivated by the way this ordering arises for Weyl groups in connection with inclusions among closures of Bruhat cells for a corresponding semisimple algebraic group. In view of the way the ordering is defined, it should not be surprising to find the Strong Exchange Condition used below in investigating its properties.

The definition has a one-sided appearance, since we have written t on the right in defining the arrow relation. But this version could just as well be replaced by a left-sided version. Say $w = w's_\alpha$, with $\ell(w) > \ell(w')$. Setting $\beta = w'(\alpha)$, we get $(w')^{-1}s_\beta w' = s_\alpha$, hence $w = s_\beta w'$. (On the other hand, if we had insisted that t belong to S, the resulting partial ordering, sometimes called the **weak ordering**, would actually have a one-sided nature, as the reader can check for dihedral groups. We won't pursue this possibility here, but see Björner [2].)

One other remark about the definition: when $w' \to w$, the precise length difference is not specified; it must be odd but need not be 1 (as seen already in dihedral groups). So it is not clear at first whether two immediately adjacent elements in the Bruhat ordering must differ in length by just 1. This turns out to be true, but requires some delicate arguments (5.11).

Another natural question about the ordering will also be deferred. If $I \subset S$, the Coxeter group W_I has a Bruhat ordering of its own; does this agree with the restriction to W_I of the Bruhat ordering of W? The answer will be given in 5.10.

Exercise. Prove that $v < w$ if and only if $v^{-1} < w^{-1}$.

Example 1. If W is a dihedral group $\mathcal{D}_m$, $m \le \infty$, all elements of distinct lengths are comparable in the Bruhat ordering (but not in the weak ordering): $v < w$ if and only if $\ell(v) < \ell(w)$.

Example 2. If W is the symmetric group $\mathcal{S}_n$, each element π can be represented by the string of n integers $(\pi(1), \ldots, \pi(n))$. Then $\pi \le \sigma$ if and only if σ is obtainable from π by a sequence of transpositions (ij), where $i < j$ and i occurs to the left of j in π. For example, when $n = 5$, we have $24153 \to 42153 \to 45123 \to 54123$, or more directly, $24153 \to 54123$. Another criterion, due to Deodhar, goes as follows. Given a sequence of integers $(a_1, \ldots, a_k)$, denote by $[a_1, \ldots, a_k]$ the sequence rewritten in increasing order. Order $\mathbf{Z}^k$ by $(a_1, \ldots, a_k) \le (b_1, \ldots, b_k)$ iff $a_i \le b_i$ for all i. Then $\pi \le \sigma$ iff $[\pi(1), \ldots, \pi(k)] \le [\sigma(1), \ldots, \sigma(k)]$ for $1 \le k \le n$.

Example 3. There is added symmetry in case W is finite, with longest element $w_\circ$ (see Exercise 2 in 5.6). One sees easily that $v \le w$ if and only if $w_\circ w \le w_\circ v$. (This will be used in 7.6.)

One rather subtle property of the Bruhat ordering is needed in 5.10:

Proposition *Let* $w' \le w$ *and* $s \in S$. *Then either* $w's \le w$ *or else* $w's \le ws$ *(or both).*

Proof. The proof reduces quickly (as the reader should check) to the case $w' \to w$, where $w = w't$ $(t \in T)$ and $\ell(w) > \ell(w')$. If $s = t$, there is nothing to prove, so we assume $s \neq t$. Two cases have to be analyzed:

(a) If $\ell(w's) = \ell(w') - 1$, then $w's \to w' \to w$, forcing $w's \leq w$.

(b) If $\ell(w's) = \ell(w') + 1$, we shall argue that $w's < ws$. Since $(w's)t' = ws$ for the reflection $t' = sts$, it is enough to show that $\ell(w's) < \ell(ws)$. Suppose the contrary, i.e., $\ell(ws) < \ell(w's)$. Then the Strong Exchange Condition (5.8) can be applied to the pair $t', w's$ as follows. For any reduced expression $w' = s_1 \cdots s_r$, $w's = s_1 \cdots s_r s$ is also reduced, since $\ell(w's) > \ell(w')$ by assumption. Then $ws = (w's)t'$ is obtained from $w's$ by omitting one factor in this reduced decomposition. This factor cannot be s, since $s \neq t$. Thus $ws = s_1 \cdots \hat{s}_i \cdots s_r s$ for some i, or $w = s_1 \cdots \hat{s}_i \cdots s_r$, contradicting $\ell(w) > \ell(w')$. $\square$

5.10   Subexpressions

There is a very simple and useful characterization of the Bruhat ordering in terms of **subexpressions** of a given reduced expression $w = s_1 s_2 \cdots s_r$, by which we mean products (not necessarily reduced, and possibly empty) of the form $s_{i_1} \cdots s_{i_q}$ $(1 \leq i_1 < i_2 < \ldots < i_q \leq r)$. Formally, the given reduced expression is an ordered r-tuple of elements of S, and a subexpression is a q-tuple obtained by discarding some or all of these elements.

Theorem   *Let $w = s_1 \cdots s_r$ be a fixed, but arbitrary, reduced expression for w. Then $w' \leq w$ if and only if w' can be obtained as a subexpression of this reduced expression.*

Proof. Let us first show that any $w' < w$ occurs as a subexpression of the given reduced expression for w. Start with the case $w' \to w$, say $w = w't$. Since $\ell(w') < \ell(w)$, the Strong Exchange Condition can be applied to the pair t, w to yield $w' = wt = s_1 \cdots \hat{s}_i \cdots s_r$ for some i. This argument can be iterated. If in turn $w'' \to w'$, with $w'' = w't'$, apply the Strong Exchange Condition to the pair $t', w' = s_1 \cdots \hat{s}_i \cdots s_r$ (which is not required to be a reduced expression!) to obtain

$$w'' = w't' = s_1 \cdots \hat{s}_i \cdots \hat{s}_j \cdots s_r$$

or else

$$w'' = s_1 \cdots \hat{s}_j \cdots \hat{s}_i \cdots s_r.$$

In the other direction, we are given a subexpression $s_{i_1} \cdots s_{i_q}$ and must show it to be $\leq w$. Here we can use induction on $r = \ell(w)$, the case $r = 0$ being trivial. If $i_q < r$, the induction hypothesis can be applied to the reduced expression $s_1 \cdots s_{r-1}$ to yield:

$$s_{i_1} \cdots s_{i_q} \leq s_1 \cdots s_{r-1} = ws_r < w.$$

If $i_q = r$ we first use induction to get $s_{i_1} \cdots s_{i_{q-1}} \leq s_1 \cdots s_{r-1}$, and then apply Proposition 5.9 to get either

$$s_{i_1} s_{i_2} \cdots s_{i_q} \leq s_1 \cdots s_{r-1} < w$$

or else

$$s_{i_1} s_{i_2} \cdots s_{i_q} \leq s_1 s_2 \cdots s_r = w. \quad \square$$

This characterization of the Bruhat ordering would be awkward to use as the initial definition, because of the apparent dependence on a fixed choice of reduced expression, e.g., transitivity would be far from obvious. But it helps to make explicit computations more transparent, as in the exercise below. And it answers a natural question about parabolic subgroups:

Corollary *If $I \subset S$, the Bruhat ordering of W agrees on W_I with the Bruhat ordering of the Coxeter group W_I.*

Proof. If $w \in W_I$, it has a reduced expression (in W) involving just elements of I, thanks to 5.5. By the theorem, the elements $\leq w$ in the Bruhat ordering of either W or W_I are the subexpressions of this reduced expression. $\square$

Exercise. Describe the Bruhat ordering of S_4, and verify that directly adjacent elements always differ in length by 1. Further verify that in each closed interval

$$[w', w] := \{x \in W | w' \leq x \leq w\},$$

the number of elements of even length equals the number of elements of odd length. For example, if we write $S = \{s_1, s_2, s_3\}$ with $s_1 s_3 = s_3 s_1$, the interval from 1 to $s_1 s_2 s_3 s_2$ contains the following intermediate elements:

$$s_1, \ s_2, \ s_3, \ s_1 s_2, \ s_1 s_3, \ s_2 s_3, \ s_3 s_2, \ s_1 s_2 s_3, \ s_1 s_3 s_2, \ s_2 s_3 s_2.$$

(For a picture, see Shi [1], p. 20, or Björner [2].)

5.11 Intervals in the Bruhat ordering

To show that elements of W directly adjacent in the Bruhat ordering must differ in length by just 1, we first examine closely a configuration which will arise in the proof.

Lemma *Let $w' < w$, with $\ell(w) = \ell(w') + 1$. Suppose there exists $s \in S$ for which $w' < w's$ (i.e., $\ell(w') < \ell(w's)$) and $w's \neq w$. Then both $w < ws$ and $w's < ws$.*

Proof. Proposition 5.9 implies that $w's \leq w$ or $w's \leq ws$. The first is impossible since the lengths are equal and $w's \neq w$. Since $w' \neq w$, we get $w's < ws$. In turn, $\ell(w) = \ell(w's) < \ell(ws)$, forcing $w < ws$. $\square$

Exercise 1. Prove a dual version of the lemma: supposing $ws < w$ and $ws \neq w'$, conclude that both $w's < w'$ and $w's < ws$.

Proposition   *Let $w' < w$. Then there exist $w_0, \ldots, w_m \in W$ such that $w' = w_0 < w_1 < \ldots < w_m = w$, and $\ell(w_i) = \ell(w_{i-1}) + 1$ for $1 \leq i \leq m$.*

Proof. Proceed by induction on $\ell(w) + \ell(w')$. If this is 1, then $w' = 1$ and $w \in S$, so there is nothing to prove. Now $w \neq 1$, so $\ell(ws) < \ell(w)$ for some $s \in S$, say $s = s_r$ in a reduced expression $w = s_1 \cdots s_r$ (from the Exchange Condition). By Theorem 5.10, $w' = s_{i_1} \cdots s_{i_q}$ for some $i_1 < \ldots < i_q$. Two cases are possible:

(a) Suppose $w' < w's$. If $i_q = r$, note that $w's$ is also a subexpression of $ws = s_1 \cdots s_{r-1} < w$, so we have $w' < w's \leq ws < w$. By induction, we can find a chain of the desired type from w' to ws, and one more step gets us to w. On the other hand, if $i_q \neq r$, then w' is itself a subexpression of $ws < w$, and induction similarly applies.

(b) Suppose $w's < w'$. Now induction provides a chain from $w's$ up to w:
$$w's = w_0 < w_1 < \ldots < w_m = w,$$
with $\ell(w_i) = \ell(w_{i-1})+1$. Choose the smallest index i for which $w_is < w_i$. Note that $w_0s = w' > w's = w_0$, while $w_ms = ws < w = w_m$, so such an $i \geq 1$ does exist. We claim that $w_i = w_{i-1}s$. Otherwise we could apply the above lemma to the situation:
$$w_{i-1} < w_{i-1}s \neq w_i$$
to get $w_i < w_is$, contrary to the choice of i. Thus $w_i = w_{i-1}s$. However, for $1 \leq j < i$, we have instead $w_j \neq w_{j-1}s$, because $w_j < w_js$. For such indices j, the lemma can be applied to the situation:
$$w_{j-1} < w_{j-1}s \neq w_j$$
to obtain $w_{j-1}s < w_js$. Combining these observations, we find a chain of the desired type:
$$w' = w_0s < w_1s < \ldots < w_{i-1}s = w_i < w_{i+1} < \ldots < w_m = w. \quad \square$$

Exercise 2. An order-preserving bijection $W \to W$ also preserves lengths. (Examples of such bijections?)

5.12   Poincaré series

In the remainder of this chapter we examine more carefully the parabolic subgroups of W, in both combinatorial and geometric settings.

First we generalize straightforwardly the description in 1.10 of a distinguished set of coset representatives for W/W_I, $I \subset S$. Define

$$W^I := \{w \in W | \ell(ws) > \ell(w) \text{ for all } s \in I\}.$$

Then the proof of part (c) of Proposition 1.10 can be repeated word-for-word to obtain the same result for an arbitrary Coxeter group:

> *Fix $I \subset S$. Given $w \in W$, there is a unique $u \in W^I$ and a unique $v \in W_I$ such that $w = uv$. Then $\ell(w) = \ell(u) + \ell(v)$. Moreover, u is the unique element of smallest length in the coset wW_I.*

With this in hand we can generalize to W the construction of Poincaré polynomials in 1.11. But when W is infinite, we get a formal power series in the indeterminate t. As before, we define

$$W(t) := \sum_{n \geq 0} a_n t^n,$$

where $a_n := \text{Card } \{w \in W | \ell(w) = n\}$. (This is finite, since S is finite.) We call $W(t)$ the **Poincaré series** of W.

As in 1.11, we can similarly define $X(t)$ for an arbitrary subset X of W, by counting only the number of elements of X of each length. In particular, $W_I(t)$ coincides with the Poincaré series of the Coxeter group W_I, since $\ell = \ell_I$ on W_I. It also follows immediately from the discussion above that

$$W(t) = W_I(t)W^I(t).$$

Before stating the analogue of Proposition 1.11, we recall Exercise 2 in 5.6: when W is infinite the length function takes arbitrarily large values whereas, when W is finite, there is a unique element w_o of maximum length N (sending all positive roots to negative roots). This is an immediate consequence of Proposition 5.6. To simplify notation, write $(-1)^I$ instead of $(-1)^{|I|}$.

Proposition (a) *In the field of formal power series in t, we have the identity*

$$\sum_{I \subset S}(-1)^I \frac{W(t)}{W_I(t)} = \sum_{I \subset S}(-1)^I W^I(t) = 0,$$

unless W is finite, in which case the right side equals t^N.

(b) *$W(t)$ is an explicitly computable rational function of t.*

Proof. (a) When W is finite, the proof of Proposition 1.11 may be repeated *verbatim*. When W is infinite, the set $K := \{s \in S | \ell(ws) > \ell(w)\}$ used in that proof is nonempty for all w, so the calculation there results in a right hand side equal to 0.

(b) Proceed by induction on $|S|$; if this is 1, then $W(t) = 1 + t$. In general, use the equation in part (a), first moving the term for which $I = S$ from the left side to the right side and then dividing both sides by $W(t)$. This yields:

$$\sum_{I \neq S} (-1)^I \frac{1}{W_I(t)} = \frac{f(t)}{W(t)},$$

where $f(t) := -(-1)^S$ unless W is finite, in which case $f(t) := t^N - (-1)^S$. The left side involves only those $W_I(t)$ for which $I \neq S$, and is therefore a computable rational function of t, by induction. Therefore $W(t)$ can also be computed as a rational function. $\square$

What is involved in carrying out the computation of $W(t)$ by this method can be illustrated quickly in the case of the infinite dihedral group $W = \mathcal{D}_\infty$, where $S = \{s, s'\}$ and $m(s, s') = \infty$. It follows directly from the definition that $W(t) = 1 + 2t + 2t^2 + \ldots$ The inductive approach to computing $W(t)$ as a rational function requires knowing that $W_I(t) = 1$ when $I = \emptyset$, while $W_I(t) = 1 + t$ when $I = \{s\}$ or $\{s'\}$. Accordingly, the left side above becomes

$$1 - \frac{1}{1+t} - \frac{1}{1+t}$$

and the right side is $-1/W(t)$. Thus we get the expected result

$$W(t) = \frac{1+t}{1-t}.$$

Exercise. Let W_a be the affine Weyl group of type $\widetilde{C_2}$, with $S = \{s_0, s_1, s_2\}$ and $m(s_0, s_1) = 4 = m(s_1, s_2)$, $m(s_0, s_2) = 2$. Show that

$$W_a(t) = \frac{(1 - t^2)(1 - t^4)}{(1 - t)^3(1 - t^3)}.$$

How is this related to the Poincaré polynomial of the corresponding Weyl group? (This and the example $\mathcal{D}_\infty$ illustrate a general theorem of Bott on affine Weyl groups (8.9).)

5.13   Fundamental domain for W

Theorem 5.4 is the key fact about how W acts on V in the geometric representation $\sigma : W \to \mathrm{GL}(V)$. To get more insight into the action of parabolic subgroups, we must make the geometry of the situation more explicit, along the lines of 1.12. But we no longer have a euclidean inner product to work with; indeed, the bilinear form B may well be degenerate. There is no direct analogue here of the positive and negative half-spaces defined by a reflecting hyperplane. To restore at least part of the analogy with 1.12, we consider the contragredient action $\sigma^* : W \to GL(V^*)$. Elements of V^* will be denoted by $f, g, h, \ldots$, and the natural pairing with V will be denoted by $\langle f, \lambda \rangle$. Then the action of W on V^* is characterized by:

$$\langle w(f), w(\lambda) \rangle = \langle f, \lambda \rangle \text{ for } w \in W, f \in V^*, \lambda \in V.$$

We can now introduce for each $s \in S$ the hyperplane

$$Z_s := \{f \in V^* | \langle f, \alpha_s \rangle = 0\},$$

together with the associated half-spaces

$$A_s := \{f \in V^* | \langle f, \alpha_s \rangle > 0\},$$

$$A'_s := \{f \in V^* | \langle f, \alpha_s \rangle < 0\} = s(A_s).$$

Finally, let C be the intersection of all A_s, $s \in S$.

Observe that s fixes Z_s pointwise. Take α_s to be the first element in an ordered basis of V, followed by a basis of the fixed point space of σ_s. Denote the dual basis by $f_1, \ldots, f_n$, $n = |S|$. Then, for all $i > 1$, the effect of $s(f_i)$ on the basis of V is clearly the same as that of f_i.

If we identify V with $\mathbf{R}^n$ $(n = |S|)$, say by fixing the basis consisting of all α_s $(s \in S)$, then V^* with the dual basis may also be identified with $\mathbf{R}^n$. Relative to the standard topology of $\mathbf{R}^n$ (which has nothing to do with the bilinear form B), Z_s is closed, while each of A_s and A'_s is open; therefore C is open. It is clear that the closure $\overline{A_s}$ of A_s is $A_s \cup Z_s$, and in turn $D := \bar{C}$ is the intersection of all $\overline{A_s}$. Note too that the action of each $w \in W$ on V or V^* is continuous. While we might avoid the use of such topological information in the theorem below, some steps would become less transparent. (And in determining the finite Coxeter groups in Chapter 6 below, the topological viewpoint will be essential.)

The object now is to study the action of the parabolic subgroups W_I by partitioning D into corresponding subsets C_I, defined by

$$C_I := \left(\bigcap_{s \in I} Z_s \right) \cap \left(\bigcap_{s \notin I} A_s \right).$$

At the extremes, $C_\emptyset = C$, while $C_S = \{0\}$. Since s fixes Z_s pointwise, W_I fixes each point of C_I. In the other direction, if $s \in S$ fixes a point $f \in C_I$, then s must belong to I: $\langle f, \alpha_s \rangle = \langle s(f), s(\alpha_s) \rangle = -\langle f, \alpha_s \rangle$ forces $f \in Z_s$. But it is not yet clear that W_I is the full stabilizer of each point.

Define U to be the union of all $w(D)$, $w \in W$. This is a W-stable subset of V^*, the union of the family $\mathcal{C}$ of all sets of the form $w(C_I)$, where $w \in W$ and $I \subset S$. It will be shown below that these sets in fact form a partition of U. Note that D is a convex cone. This implies at once that U is a cone: the **Tits cone**. It will be seen below that U is also convex. The action of W on the Tits cone is what we can describe rather well. (However, it is easy to see that U is a proper subset of V^* unless W is finite; see the exercise below.)

Lemma Let $s \in S$ and $w \in W$. Then $\ell(sw) > \ell(w)$ if and only if $w(C) \subset A_s$, whereas $\ell(sw) < \ell(w)$ if and only if $w(C) \subset A'_s$.

Proof. This is just a translation of Theorem 5.4. Indeed, $\ell(sw) > \ell(w)$ means that $\ell(w^{-1}s) > \ell(w^{-1})$, which is equivalent to $w^{-1}(\alpha_s) > 0$. If $f \in C$, then $\langle w(f), \alpha_s \rangle > 0$ means that $\langle f, w^{-1}(\alpha_s) \rangle > 0$, which is equivalent (by the way C is defined) to saying that $w^{-1}(\alpha_s) > 0$. So $w(C) \subset A_s$ if and only if $\ell(sw) > \ell(w)$. $\square$

Theorem (a) Let $w \in W$ and $I, J \subset S$. If $w(C_I) \cap C_J \neq \emptyset$, then $I = J$ and $w \in W_I$, so $w(C_I) = C_I$. In particular, W_I is the precise stabilizer in W of each point of C_I, and C is a partition of U.

(b) D is a fundamental domain for the action of W on U: the W-orbit of each point of U meets D in exactly one point.

(c) The cone U is convex, and every closed line segment in U meets just finitely many of the sets in the family C.

Proof. (a) Proceed by induction on $\ell(w)$, the case $w = 1$ being obvious. If $\ell(w) > 0$, write $w = s(sw)$ with $\ell(sw) < \ell(w)$ for some $s \in S$. The lemma above forces $w(C) \subset s(A_s) = A'_s$, whence by continuity $w(D) \subset \overline{A'_s}$. Combined with the fact that $D \subset \overline{A_s}$, this implies that $D \cap w(D) \subset Z_s$. Thus s fixes each point in the intersection – in particular, each point in the (nonempty!) set $C_J \cap w(C_I)$. Two things follow. First, s fixes some point of C_J and hence (as remarked earlier) $s \in J$. Second, $C_J \cap sw(C_I) = s(C_J \cap w(C_I))$ is nonempty. The induction hypothesis, applied to sw, shows that $I = J$ and $sw \in W_I$. Since $s \in J = I$, we get $w \in W_I$ as required. We conclude that the sets $w(C_I)$ comprising the family C are all disjoint, as w runs over coset representatives in W/W_I and I runs over the subsets of S.

(b) By definition of U, each W-orbit in U meets D in at least one point. Suppose $f, g \in D$ both lie in the same W-orbit: $w(f) = g$ for some $w \in W$. Say $f \in C_I, g \in C_J$, so that $w(C_I) \cap C_J$ is nonempty. By part (a), $I = J$ and $w \in W_I$, forcing $f = w(f) = g$.

(c) It is enough to prove: *if $f, g \in U$, the closed segment $[f\ g]$ joining them is covered by finitely many of the sets in C.*

This is clear when both f and g belong to D, which is convex and is covered by the sets C_I. In general we may replace f, g by their images under some element of W; so without loss of generality we may assume that $f \in D$ and $g \in w(D)$. Now proceed by induction on $\ell(w)$, the case $w = 1$ having just been dealt with. Let $\ell(w) > 0$. The segment $[f\ g]$

intersects D in a closed segment $[f\ h]$, which can be covered by finitely many sets from $\mathcal{C}$. It remains to cover $[h\ g]$. We may assume $g \notin D$. Say $g \in A'_s$ for $s \in I$ and $g \in \overline{A_s}$ for $s \notin I$. If we had $h \in A_s$ for all $s \in I$, then all nearby points k on $[h\ g]$ would also satisfy $k \in A_s\ (s \in I), k \in \overline{A_s}\ (s \notin I)$ and hence lie in D, which is absurd. Therefore $h \in Z_s$ for some $s \in I$. Since $g \in A'_s$, we must have $w(D) \subset \overline{A'_s}$, hence $w(C) \subset A'_s$. By the above lemma, $\ell(sw) < \ell(w)$. So the induction hypothesis may be applied to $h \in D$ and $s(g) \in sw(D)$. Thus the segment from h to $s(g)$ has a finite cover from $\mathcal{C}$, and transforming the picture by s yields a finite cover of the segment from $s(h) = h$ to $s^2(g) = g$. $\square$

Exercise. If the Tits cone U is equal to V^*, prove that W is finite. [Find $w \in W$ for which $w(\bar{C})$ meets $-C$. Then show that $w^{-1}(\alpha_s) < 0$ for all $s \in S$, and deduce that W is finite.] Conversely, if W is finite, it will be seen in 6.4 that V^* is a euclidean space, with W acting as in Chapter 1; thus $U = V^*$ will follow.

Remark. The theorem provides a concrete (though rather impractical) way to solve the **Word Problem** for W. Fix the basis of V consisting of the $\alpha_s\ (s \in S)$ and let $\{f_s | s \in S\}$ be the dual basis. Then $f := \sum f_s$ lies in C, and is therefore fixed by no element of W except 1. To decide whether or not a given product of elements from S is equal to 1 in W, apply the corresponding product of elements $\sigma^*(s)$ to f and see whether or not the result is f. (See 8.1 for another approach to the Word Problem.)

As in 1.15, we can formulate the geometry here as a **Coxeter complex**, based on the family of parabolic subgroups of W. It can be regarded as an abstract simplicial complex, and provides an essential ingredient in the more elaborate complexes introduced by Tits and known as **buildings**. For a thorough account of all this, see Brown [1], Ronan [1], Tits [6].

Notes

The basic facts about Coxeter groups are developed in Bourbaki [1], IV, §1, following earlier work of Coxeter, Witt, Tits, and others.

(5.3) The exercise is based on Bourbaki [1], IV, 1.3, Prop. 3.

(5.4)–(5.7) Our treatment is heavily influenced by Deodhar [4][7], who developed the general notion of root system for a Coxeter group. The proof of Theorem 5.5 follows suggestions of E. Neher.

(5.8) Verma [2] introduced the Strong Exchange Condition. Exercise 4 here is due to him. We follow Deodhar [4] (from which Exercise 3 is drawn).

(5.9)–(5.11) The Bruhat ordering of a Weyl group had been implicit in the literature for some time, notably in Chevalley's study of Schubert varieties, before its formal development in Verma's 1966 thesis and Verma [1], Steinberg [4], Bernstein–Gelfand–Gelfand [2], Deodhar [1][2][5][7][8][9]. See also Björner [2], Hiller [3], Proctor [1], Shi [1], Stanley [3], and (8.5)–(8.8) below. While it would be more apt historically to adopt the terminology 'Chevalley ordering' (as pointed out recently by A. Borel), there is by now a large amount of literature referring to the 'Bruhat ordering'.

(5.12) See Steinberg [5], 1.25.

(5.13) We follow Bourbaki [1], V, 4.6.

Chapter 6

Special cases

Having laid out the general theory of Coxeter groups in Chapter 5, we turn our attention to some of the most important special cases. First we indicate how to reduce most questions to the case when the Coxeter graph is connected (6.1). Then we reconsider the geometric representation of W relative to the bilinear form B, and show in 6.4 that the only finite Coxeter groups are the finite reflection groups studied in Part I. The treatment follows Bourbaki [1], V, §4, based on Witt [1].

We also compare in 6.5 the geometric representation of an affine Weyl group with the description given in Chapter 4. In 6.6 we characterize 'crystallographic' Coxeter groups. Another large class of interesting examples consists of 'hyperbolic Coxeter groups' (6.8).

As a matter of notation, we use both s and t to denote elements of S, when (W, S) is a Coxeter system.

6.1 Irreducible Coxeter systems

We say a Coxeter system (W, S) is **irreducible** if the Coxeter graph Γ is connected, as in 2.2. The argument of Proposition 2.2 can be repeated here:

Proposition *Let (W, S) be any Coxeter system. If $\Gamma_1, \ldots, \Gamma_r$ are the connected components of the Coxeter graph Γ, let $S_1, \ldots, S_r$ be the corresponding subsets of S. Then W is the direct product of the parabolic subgroups $W_{S_1}, \ldots, W_{S_r}$, and each Coxeter system (W_{S_i}, S_i) is irreducible.*

Proof. Use induction on r. Since the elements of S_i commute with the elements of S_j when $i \neq j$, it is clear that the indicated parabolic subgroups centralize each other, hence that each is normal in W. Moreover, the product of these subgroups contains S and therefore must be all of W. By induction, $W_{S \setminus S_i}$ is the direct product of the remaining W_{S_j}, and part (c) of Corollary 5.10 implies that W_{S_i} intersects it trivially. So the product is direct. $\square$

6.2   More on the geometric representation

Recall from 5.3 the construction of the representation $\sigma : W \to \mathrm{GL}(V)$, which was shown in 5.4 to be faithful. The elements of S are represented by 'reflections' σ_s relative to the bilinear form B on V. On the basis $\{\alpha_s | s \in S\}$, B is defined by:

$$B(\alpha_s, \alpha_t) = -\cos \frac{\pi}{m(s,t)}.$$

We need to examine more closely some topological features of this situation, based on the discussion in 5.13. Relative to any fixed ordered basis of V, we may identify V with $\mathbf{R}^n$ and $\mathrm{GL}(V)$ with $\mathrm{GL}(n, \mathbf{R})$, the latter in turn being viewed as a subspace of $\mathbf{R}^{n^2}$ in an obvious way. Note that $\mathrm{GL}(n, \mathbf{R})$ is an open set, being the set of non-zeros of the determinant polynomial on the set of all $n \times n$ matrices. Note too that multiplication of a set of matrices by one fixed matrix induces a homeomorphism of the space of $n \times n$ matrices.

By using a dual basis for V^*, we get similar identifications for V^* and $\mathrm{GL}(V^*)$. Recall from 5.13 the open set C in V^* (an intersection of finitely many open half-spaces), whose closure is a fundamental domain for the action of W on the union of all its W-translates. It is clear that, for any fixed $f \in V^*$, the orbit map $\mathrm{GL}(V^*) \to V^*$ sending $g \mapsto g \cdot f$ is continuous (being given in coordinate form by linear polynomials). Thus the inverse image of C (call it C_0) is an open neighborhood of the identity element 1 in $\mathrm{GL}(V^*)$. Choose $f \in C$. Then Theorem 5.13 implies that $\sigma^*(W) \cap C_0 = \{1\}$. In turn, an arbitrary element $g = \sigma^*(w)$ has an open neighborhood gC_0 intersecting $\sigma^*(W)$ in $\{g\}$. This means that $\sigma^*(W)$ is a *discrete* subset of $\mathrm{GL}(V^*)$. By 'transport of structure' we obtain:

Proposition $\sigma(W)$ *is a discrete subgroup of* $\mathrm{GL}(V)$, *topologized as above.* $\square$

Corollary   *If the form B is positive definite, then W is finite.*

Proof. If B is positive definite, V is just a euclidean space, which may be identified with V^*. By using an orthonormal basis of V in the discussion above, we identify $\sigma(W)$ with a subgroup of the orthogonal group $\mathrm{O}(n, \mathbf{R}) \subset \mathrm{GL}(n, \mathbf{R})$. It is well-known that $\mathrm{O}(n, \mathbf{R})$ is a compact subset of the set of all $n \times n$ matrices : it is closed by virtue of being defined by polynomial equations (matrix times transpose equals 1), and it is bounded because the rows (or columns) of an orthogonal matrix are unit vectors. On the other hand, the proposition shows that $\sigma(W)$ is a discrete subgroup of $O(n, \mathbf{R})$. Since a discrete subgroup of a compact Hausdorff group is closed (hence finite), $W \cong \sigma(W)$ is finite. □

We shall prove the converse of this corollary in 6.4 below, thus showing that finite Coxeter groups are the same as finite reflection groups. The characterization in terms of positive definiteness of B of course fits in well with the strategy we adopted in Chapter 2 for classifying finite reflection groups.

6.3 Radical of the bilinear form

As illustrated by dihedral groups (finite and infinite), the symmetric bilinear form B on V may or may not be nondegenerate. Here we look more closely at the radical of B:

$$V^{\perp} := \{\lambda \in V \,|\, B(\lambda, \mu) = 0 \text{ for all } \mu \in V\}.$$

Note first of all that $V^{\perp}$ is a W-invariant proper subspace, since B is W-invariant and not identically 0. We claim that $V^{\perp} = \bigcap_{s \in S} H_s$ (where H_s is the orthogonal complement of α_s relative to B) and is therefore fixed pointwise by W. One inclusion is clear. In the other direction, $B(\lambda, \alpha_s) = 0$ for all $s \in S$ forces $\lambda \in V^{\perp}$, since the α_s span V.

Proposition *Assume that (W, S) is irreducible.*

(a) *Every proper W-invariant subspace of V is included in the radical $V^{\perp}$ of the form B, where $V^{\perp} = \bigcap_{s \in S} H_s$ is fixed pointwise by W.*

(b) *If B is degenerate, then V fails to be completely reducible as a W-module.*

(c) *If B is nondegenerate, then V is irreducible as a W-module.*

(d) *The only endomorphisms of V commuting with the action of W are the scalars.*

Proof. (a) Let $V' \neq V$ be a W-invariant subspace. Suppose first that no root α_s ($s \in S$) lies in V'. Each σ_s acts semisimply on V' (with possible eigenvalues $1, -1$). But the (-1)-eigenspace (spanned by α_s) does not occur in V', by assumption. This forces σ_s to fix V' pointwise, so that V' lies in the intersection of all H_s, which has been observed to be $V^{\perp}$.

What happens if some α_s does lie in V'? Take any neighbor t of s in the Coxeter graph Γ, so that $\sigma_t(\alpha_s) = \alpha_s + c\alpha_t$ for some nonzero c. Since $\sigma_t(\alpha_s) \in V'$, this forces $\alpha_t \in V'$ as well. But Γ is connected, so we can proceed step-by-step to get all α_t ($t \in S$) in V', whence $V' = V$ contrary to hypothesis.

(b) If B is degenerate, $V^{\perp}$ is a proper nonzero W-invariant subspace, which according to part (a) cannot have any W-invariant complement.

(c) If B is nondegenerate, part (a) implies that V has no nonzero proper W-submodules.

(d) Suppose an endomorphism z of V commutes with all $\sigma(w)$, $w \in W$. Fix any $s \in S$. Since z commutes with σ_s, the line L spanned by α_s is z-invariant, so z acts there with an eigenvalue c. We claim that z is just c times the identity operator 1. Consider the kernel V' of $t - c \cdot 1$. This is clearly stable under $\sigma(W)$, and contains L, which does not lie in $V^\perp$. Thanks to part (a), we must have $V' = V$. $\square$

Exercise 1. Assume that (W, S) is irreducible. Determine the center $Z(W)$ of W as follows. If W is finite and $w_\circ = -1$ (where $w_\circ$ is the unique longest element, as in Exercise 2 of 5.6), then $Z(W) = \{1, -1\}$. Otherwise $Z(W) = \{1\}$.

Exercise 2. Suppose $\rho : G \to \mathrm{GL}(E)$ is an irreducible representation of a group G, with dim $E < \infty$. If G contains at least one pseudo-reflection (an element of finite order whose 1-eigenspace has codimension 1 in E), then the only endomorphisms of E commuting with $\rho(G)$ are the scalars.

6.4 Finite Coxeter groups

Our goal is to show that the finite Coxeter groups are precisely the finite reflection groups studied in Chapter 1. For this we need to review some standard facts about group representations.

Lemma *Let* $\rho : G \to \mathrm{GL}(E)$ *be a group representation, with* E *a finite dimensional vector space over* $\mathbf{R}$.

(a) *If* G *is finite, then there exists a positive definite G-invariant bilinear form on* E.

(b) *If* G *is finite, then* ρ *is completely reducible.*

(c) *Suppose the only endomorphisms of* E *commuting with* $\rho(G)$ *are the scalars. If* β *and* β' *are nondegenerate symmetric bilinear forms on* E, *both G-invariant, then* β' *is a scalar multiple of* β.

Proof. (a) Start with any positive definite symmetric bilinear form β on E, and 'average' it over G to obtain one which is also G-invariant:

$$\bar{\beta}(\lambda, \mu) := \sum_{g \in G} \beta(g \cdot \lambda, g \cdot \mu),$$

where $\lambda, \mu \in E$ and $g \cdot \lambda = \rho(g)(\lambda)$, etc.

(b) Now E is the direct sum of any subspace and its orthogonal complement relative to the positive definite form $\bar{\beta}$ constructed in (a), by nondegeneracy. On the other hand, the orthogonal complement of a G-invariant subspace is also G-invariant, since $\bar{\beta}$ is an invariant form. Complete reducibility follows.

(c) Any nondegenerate form sets up a vector space isomorphism between E and its dual space E^* in the usual way. When the form is invariant, this becomes an isomorphism of G-modules (relative to ρ and its contragredient). Composing the isomorphism defined by β with the inverse of that defined by β' gives a G-module isomorphism of E onto itself, i.e., an endomorphism of E commuting with $\rho(G)$. By assumption, this is just a scalar, so β and β' are proportional. $\square$

Theorem *The following conditions on the Coxeter group W are equivalent:*

(a) *W is finite.*

(b) *The bilinear form B is positive definite.*

(c) *W is a finite reflection group (in the sense of Chapter 1).*

Proof. Without loss of generality, we may assume that (W, S) is an irreducible Coxeter system.

(a) $\Rightarrow$ (b). Thanks to part (b) of the lemma above, W acts completely reducibly on V. Then part (b) of Proposition 6.3 implies that B must be nondegenerate. In turn, part (c) of that proposition says that W acts irreducibly, and (d) says that the scalars are the only endomorphisms of V commuting with the action of W. From part (c) of the lemma above, we conclude that (up to scalar multiples) B is the unique nondegenerate, W-invariant symmetric bilinear form on V. But, by part (a) of the lemma, there exists a positive definite W-invariant form on V, say B'. So $B' = cB$ for some nonzero $c \in \mathbf{R}$. Since $B(\alpha_s, \alpha_s) = 1$, we must have $c > 0$. Therefore B is also positive definite.

(b) $\Rightarrow$ (c). Apply Corollary 6.2.

(c) $\Rightarrow$ (a). This is immediate. $\square$

Note that when W is a finite subgroup of $\mathrm{GL}(V)$ generated by reflections (V euclidean), its geometric representation as a Coxeter group (5.3) looks just like its given representation on V: the angles between simple roots (in the sense of Chapter 1) agree with the angles between the α_s relative to B.

6.5 Affine Coxeter groups

We found in 4.7 that the Coxeter graphs of (irreducible) affine Weyl groups are precisely the positive semidefinite ones which are not positive definite. These graphs were completely determined in Chapter 2.

The affine Weyl group W_a was constructed in Chapter 4 as a group generated by affine reflections in a euclidean space. On the other hand, the geometric representation of W_a as a Coxeter group (5.3) provides another concrete realization. In this section we discuss briefly how the two constructions are related.

Begin with the irreducible Coxeter system (W, S) of rank n whose graph is one of those in Figure 2 of 2.5; so $n \geq 2$. Let $\sigma : W \to \mathrm{GL}(V)$ be its geometric representation, with associated bilinear form B and matrix A relative to the basis $(\alpha_s), s \in S$. Since A is indecomposable and positive semidefinite (but not positive definite), Proposition 2.6 shows that its nullspace is one-dimensional, spanned by a vector with strictly positive coordinates $c_s, s \in S$. Moreover, the corresponding vector $\lambda :=$ $\sum c_s \alpha_s$ spans the radical $V^\perp$ of B. The quotient space $V/V^\perp$ becomes a euclidean space (of dimension $n - 1$) relative to the positive definite form induced by B.

By Proposition 6.3, $V^\perp$ is the intersection of the hyperplanes H_s and is therefore fixed pointwise by W, i.e., $w(\lambda) = \lambda$ for all $w \in W$. As a result, W leaves stable the hyperplane orthogonal to $V^\perp$ in the dual space V^*:

$$Z := \{f \in V^* | \langle f, \lambda \rangle = 0\}.$$

As in 5.13, we write $\langle f, \lambda \rangle$ for $f(\lambda)$. Moreover, Z identifies naturally with the dual space of $V/V^\perp$ and thus acquires the structure of a euclidean space.

Note that W also stabilizes the translated affine hyperplane

$$E := \{f \in V^* | \langle f, \lambda \rangle = 1\}.$$

The euclidean structure on Z transfers naturally to E, making it an affine (euclidean) space with translation group Z. It is clear that E (like any other affine hyperplane not containing 0) spans the vector space V^*, so the action of W on E is *faithful*. Because of the euclidean structure, the isotropy group in W of any point in E acts on E as a group of *orthogonal* transformations.

Now consider how the various hyperplanes

$$Z_s := \{f \in V^* | \langle f, \alpha_s \rangle = 0\}$$

intersect E. Since $|S| > 1$ and all $c_s > 0$, λ is not proportional to any α_s and thus $Z \neq Z_s$. This forces Z_s to intersect E; the intersection is an affine hyperplane E_s in E (fixed pointwise by s). Since s acts on E as a transformation of order 2, it acts as an orthogonal reflection relative to E_s. Thus we have realized W as a subgroup of $\mathrm{GL}(E)$ generated by affine reflections.

Unless we are in type $\widetilde{A}_1$, pictured in Figure 1, all $m(s, t) < \infty$. Since st acts on E as a transformation of order $m(s, t)$, the angle between the hyperplanes E_s and E_t must agree with the corresponding angle in the realization of W as an affine Weyl group in Chapter 4. So the affine hyperplanes E_s yield the same geometric configuration as the hyperplanes bounding the alcove A_o: indeed, A_o corresponds to the region in E obtained by intersecting the cone C of 5.13 with the positive half-spaces of the various E_s. This recovers the geometric description of the affine Weyl group given in Chapter 4.

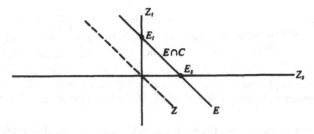

Figure 1: The case $\widetilde{A_1}$

6.6 Crystallographic Coxeter groups

In 2.8 we determined which of the finite reflection groups (acting in a euclidean space V) can stabilize a lattice in V. The answer turned out to be very simple: all integers $m(s,t)$, $s \neq t$ in S, must take one of the values 2, 3, 4, 6. We want to ask a similar question about arbitrary Coxeter groups, relative to the geometric representation $\sigma : W \to \mathrm{GL}(V)$. Call W **crystallographic relative to** σ if W stabilizes a lattice L in V.

It should be emphasized that this definition depends on the way we have chosen to represent W as a group generated by 'reflections'. There may well be other interesting representations for special classes of Coxeter groups (see the remark below). We are following Bourbaki [1], V, §4, exercise 6.

Some of the reasoning used in the finite case can be adapted easily to the general case, to yield a necessary condition for W to be crystallographic, based on the fact that the trace of $\sigma(w)$ must then lie in $\mathbf{Z}$ for all $w \in W$. Recall from 5.3 that, for $s \neq t$ in S, the corresponding reflections σ_s and σ_t generate a dihedral subgroup acting on a plane and leaving pointwise fixed its orthogonal complement relative to B. The product $\sigma_s \sigma_t$ acts on the plane as a rotation through $2\pi/m(s,t)$, if $m(s,t) < \infty$, from which we deduce as in the finite case that $m(s,t) \in \{2,3,4,6\}$. If $m(s,t) = \infty$, the trace of $\sigma_s \sigma_t$ is $n = \dim V$. So a crystallographic group must at least satisfy $m(s,t) \in \{2,3,4,6,\infty\}$ for all $s \neq t \in S$.

Unfortunately, this simple condition is not always sufficient to insure the existence of a lattice stabilized by W, in case the Coxeter graph Γ contains a circuit. If there is a circuit, label its vertices consecutively by $s_1, \ldots, s_r$ (with corresponding roots $\alpha_1, \ldots, \alpha_r$) and consider $w = s_1 \cdots s_r$. If $s = s_i$, write $\sigma_s(\alpha_j) = \alpha_j + b_{ij}\alpha_i$, so $b_{ij} = -2B(\alpha_i, \alpha_j) = 2\cos(\pi/m(s_i, s_j))$. Assuming that all $m(s,t) \in \{2,3,4,6,\infty\}$ when $s \neq t$, the respective values of b_{ij} are $0, 1, \sqrt{2}, \sqrt{3}, 2$.

To compute the trace of $\sigma(w)$, note that $\sigma(w)(\lambda) \equiv \lambda$ modulo the span of the roots $\alpha_1, \ldots, \alpha_r$, so we need only consider how $\sigma(w)$ acts there. A direct calculation shows that $\sigma(w)(\alpha_1) \equiv c\alpha_1$ modulo the span of the other α_i, where $c = b_{12}^2 + b_{r1}^2 - 1 + b_{12}b_{23} \cdots b_{r-1,r}b_{r1}$. On the other hand, for $i = 2, \ldots, r - 1$, we get $\sigma(w)(\alpha_i) \equiv (b_{i,i+1}^2 - 1)\alpha_i$ modulo the span of the other α_j, and $\sigma(w)(\alpha_r) \equiv -\alpha_r$ modulo the span of $\alpha_1, \ldots, \alpha_{r-1}$. Since the squares all lie in $\mathbf{Z}$, it follows that the trace of $\sigma(w)$ lies in $\mathbf{Z}$ if and only if $b_{12}b_{23} \cdots b_{r1} \in \mathbf{Z}$. This is true if and only if the number of edges of the circuit labelled 4 (resp. 6) is *even*. This gives another necessary condition for W to be crystallographic (relative to σ).

To see whether these necessary conditions are actually sufficient, we ask how we might construct a suitable lattice in V. The easiest procedure would be to choose a basis for the lattice consisting of vectors $\lambda_s = c_s \alpha_s, s \in S$. Suppose we are able to find scalars c_s satisfying the following conditions:

$$\begin{aligned}
m(s, t) &= 3 &&\Rightarrow && c_s = c_t \\
m(s, t) &= 4 &&\Rightarrow && c_s = \sqrt{2}c_t \text{ or } c_t = \sqrt{2}c_s \\
m(s, t) &= 6 &&\Rightarrow && c_s = \sqrt{3}c_t \text{ or } c_t = \sqrt{3}c_s \\
m(s, t) &= \infty &&\Rightarrow && c_s = c_t
\end{aligned}$$

Then we see at once that $\sigma_s(\lambda_t) = \lambda_t + d(s, t)\lambda_s$ for some $d(s, t) \in \mathbf{Z}$, whence W stabilizes the lattice with basis $\lambda_s, s \in S$.

The only problem is to see that consistent choices of the c_s can be made, under our assumptions on $m(s, t)$. If Γ contains no circuit this is straightforward: start at a terminal vertex s (having only one adjacent vertex), and choose c_s to meet the above conditions, assuming the other c_t already chosen (by induction on $|S|$).

If Γ contains one or more circuits, we have to invoke the evenness assumption on the number of edges of each circuit labelled 4 or 6. The idea is to fix an arbitrary vertex s in Γ, for which c_s is defined to be 1. Then assign values c_t by the following algorithm. Take a path of minimal length from s to t, and choose successive values of $c_{s'}$ along the path by the rule: keep the previous value if the edge is labelled 3 or ∞, but alternate multiplication and division of the previous value by $\sqrt{2}$ (resp. $\sqrt{3}$) as successive edges labelled 4 (resp. 6) are encountered. Eventually c_t is defined. This is ambiguous only if distinct minimal paths from s to t exist, but combining two such paths makes a circuit. The evenness assumption is just what is needed to make c_t well defined in this case.

Proposition *In case the Coxeter graph of W contains no circuit, W is crystallographic (relative to σ) if and only if $m(s, t) \in \{2, 3, 4, 6, \infty\}$ for all $s \neq t$ in S. Otherwise W is crystallographic (relative to σ) if and only if the same condition is fulfilled and, moreover, for each circuit in the Coxeter graph, the number of edges labelled 4 (resp. 6) is even.* $\square$

Exercise 1. If W is crystallographic relative to σ, then every parabolic subgroup W_I ($I \subset S$) is also crystallographic (relative to its geometric representation).

Exercise 2. Multiply the matrix of B by 2 to obtain a matrix A; for a parabolic subgroup W_I, denote by A_I the similarly defined matrix. Then W is crystallographic relative to σ if and only if det $A_I \in \mathbf{Z}$ for all $I \subset S$. (See Monson [1].)

Remark. In the literature the notion of **crystallographic Coxeter group** is usually defined to mean simply that all $m(s,t) \in \{2,3,4,6,\infty\}$, for $s \neq t$. One reason for this is the fact that such Coxeter groups are precisely the 'Weyl groups' of Kac–Moody Lie algebras; see Kac [1], Chapter 3. These Lie algebras are defined by generators and relations, starting with a 'generalized Cartan matrix' (a_{ij}), whose entries are integers subject to the requirements: $a_{ii} = 2, a_{ij} \leq 0$ if $i \neq j$, $a_{ij} = 0$ if and only if $a_{ji} = 0$. Such a Lie algebra has a 'root system' and a corresponding 'Weyl group' W, which stabilizes the root lattice. W turns out to be a Coxeter group on a set of generators indexed by the same index set as the matrix. The resulting values $m(s,t)$ may be labelled m_{ij} and are correlated with the a_{ij} as follows: $m_{ij} = 2,3,4,6,\infty$ when $a_{ij}a_{ji} = 0,1,2,3, \geq 4$ (respectively), $i \neq j$. (Many generalized Cartan matrices may lead to the same group W.)

6.7 Coxeter groups of rank 3

Having classified the Coxeter graphs of positive type (2.7), we see that there are many other connected graphs of rank 3. These in fact have a unified characterization, which will be a first step toward the discussion of 'hyperbolic' Coxeter groups in 6.8 below.

Let Γ be a connected Coxeter graph of rank 3. Two cases should be distinguished. First, suppose Γ is not a cycle, and label its two edges by $m, n \geq 3$:

$$\circ \overset{m}{\text{---}} \circ \overset{n}{\text{---}} \circ$$

Set $a := \cos(\pi/m), b := \cos(\pi/n)$. The matrix of the form B is then

$$\begin{pmatrix} 1 & -a & 0 \\ -a & 1 & -b \\ 0 & -b & 1 \end{pmatrix}$$

Its characteristic polynomial is $(t-1)(t^2 - 2t + c)$, where $c = 1 - a^2 - b^2$. Thus the eigenvalues are $1, 1 \pm \sqrt{a^2 + b^2}$.

These are all positive just when $a^2 + b^2 < 1$. Since the cosines in question range in value from $1/2$ to 1, there are only three such possibilities: $(m, n) = (3,3), (3,4), (3,5)$. These correspond respectively to the finite groups of types A_3, B_3, H_3.

The eigenvalue 0 occurs only when $a^2 + b^2 = 1$, which can happen in just two cases: $(m, n) = (4, 4), (3, 6)$, corresponding to the affine Weyl groups of types $\widetilde{B}_2, \widetilde{G}_2$.

In all other cases, $a^2 + b^2 > 1$, so precisely one eigenvalue is negative, and B has signature $(2,1)$.

Consider now what happens when Γ is a cycle. Label the edges $m, n, p \geq 3$, with a and b as before and with $c := \cos(\pi/p)$. Now the matrix of B is

$$\begin{pmatrix} 1 & -a & -c \\ -a & 1 & -b \\ -c & -b & 1 \end{pmatrix}$$

The determinant is $d = 1 - a^2 - b^2 - c^2 - 2abc \leq 0$, since $a, b, c \geq 1/2$, and $d = 0$ just when $a = b = c = 1/2$, or $m = n = p = 3$, corresponding to the affine Weyl group of type $\widetilde{A}_2$. Suppose on the other hand that $d < 0$. Since the trace is 3, not all eigenvalues are negative. Therefore the signature must be $(2,1)$.

Thus all but the finite and affine types lead to the same signature $(2,1)$.

There is a suggestive way to summarize the possibilities just discussed. Denote by c the sum of the reciprocals of the three labels $m(s, t), s \neq t$. Then $c > 1$ if and only if B is positive definite, $c = 1$ if and only if B is positive semidefinite (but not positive definite), $c < 1$ if and only if B is nondegenerate of signature $(2,1)$. (Here $c\pi$ can be interpreted as the sum of angles of a triangle in a geometry which is respectively spherical, euclidean, hyperbolic.)

6.8 Hyperbolic Coxeter groups

We concentrate now on the case when (W, S) is irreducible and the form B is nondegenerate, allowing us to identify V with its dual. As in 6.2, topological concepts such as connectedness come from an identification of V with $\mathbf{R}^n$ relative to some (hence any) fixed basis.

Denote by ω_s ($s \in S$) the basis dual to the basis α_s ($s \in S$), relative to B. Recall from 5.13 the cone C (now viewed as a subset of V):

$$C = \{\lambda \in V | B(\lambda, \alpha_s) > 0 \text{ for all } s \in S\} = \{\textstyle\sum c_s \omega_s | c_s > 0\}.$$

In particular, all ω_s lie in the closure D of C, which is a fundamental domain for the action of W on the union of all $w(C)$, $w \in W$. Note that D is the convex hull of the vectors ω_s.

Define the (irreducible) Coxeter system (W, S) to be **hyperbolic** if B has signature $(n-1, 1)$ and $B(\lambda, \lambda) < 0$ for all $\lambda \in C$. We also say that W is hyperbolic. (The motivation for this terminology will be discussed below.) Note that the definition forces $B(\lambda, \lambda) \leq 0$ for all $\lambda \in D$; this applies in particular to the dual basis elements ω_s.

Our aim is to characterize the corresponding Coxeter graphs in a way which will make it reasonable to carry out a complete classification. The following general lemma will be helpful.

Lemma   *Let E be an n-dimensional real vector space, endowed with a symmetric bilinear form B of signature $(n-1, 1)$. Fix a nonzero vector $\lambda \in E$, and set $H := \{\mu \in E \mid B(\lambda, \mu) = 0\}$. Then the restriction of B to H is of positive type if and only if $B(\lambda, \lambda) \leq 0$.*

Proof. Note that, if $B(\lambda, \lambda) \neq 0$, so $\lambda \notin H$, E is the orthogonal direct sum of H and the line through λ. In this case the restriction of B to H is nondegenerate, and is of positive type (in fact, positive definite) precisely when $B(\lambda, \lambda) < 0$ because of the hypothesis on the signature.

Suppose the restriction of B to H is of positive type. The preceding remarks imply that $B(\lambda, \lambda)$ cannot be strictly positive.

Conversely, suppose $B(\lambda, \lambda) \leq 0$. If $B(\lambda, \lambda) < 0$, the above remarks already show that the restriction of B to H is of positive type. If $B(\lambda, \lambda) = 0$, so $\lambda \in H$, the restriction of B to H is degenerate. But since B has signature $(n-1, 1)$, its restriction to some hyperplane H' is positive definite, and the same is true of its restriction to the hyperplane $H' \cap H$ in H (not containing λ). It follows that B is of positive type on H. $\square$

Proposition   *Let (W, S) be an irreducible Coxeter system, with graph Γ and associated bilinear form B. It is hyperbolic if and only if the following conditions are satisfied:*

(a) *B is nondegenerate, but not positive definite.*

(b) *For each $s \in S$, the Coxeter graph obtained by removing s from Γ is of positive type.*

Proof. Suppose first that W is hyperbolic, so (a) follows from the assumption on the signature of B. To verify (b), fix $s \in S$. As remarked above, $B(\omega_s, \omega_s) \leq 0$. Let L_s be the hyperplane orthogonal to ω_s. The α_t ($t \neq s$) form a basis of L_s. The lemma shows that the restriction of B to L_s is of positive type. But the matrix of this restricted form is the matrix associated with the Coxeter graph obtained by removing s from Γ, so (b) follows.

Conversely, suppose W satisfies (a) and (b). Thanks to (a), the set $N := \{\lambda \in V \mid B(\lambda, \lambda) < 0\}$ is nonempty. Because of (b), the intersection of N with each hyperplane L_s is empty, so each connected component of N lies in one of the connected components of the complement of $\bigcup_{s \in S} L_s$. These are sets of the form $\{\sum_{s \in S} c_s \alpha_s\}$, with $c_s > 0$ for certain s, and $c_s < 0$ for the others. Now we can see why the signature of B must be $(n-1, 1)$. Otherwise, owing to (a), we would have at least a two-dimensional subspace Z of V for which $Z \setminus \{0\} \subset N$. But Z is *connected* and therefore lies in one of the sets just described, which contradicts the fact that Z is closed under taking negatives.

Finally, we show that C lies in N, which is now a standard cone (minus the origin) with two connected components, each of them convex. According to (b), B is of positive type on the hyperplane L_s orthogonal to ω_s, and therefore the above lemma shows that $B(\omega_s, \omega_s) \leq 0$. So each ω_s lies in the closure of N. It follows that their convex hull D lies in the closure of N, so $C \subset N$. □

Exercise. Hyperbolic Coxeter groups can only exist in ranks ≥ 3. In rank 3, all connected Coxeter graphs which are not of positive type yield hyperbolic Coxeter groups (see the discussion in 6.7 above). Determine the hyperbolic Coxeter groups of rank 4.

As promised above, we discuss briefly some of the motivation behind the definition of hyperbolic Coxeter groups. See Bourbaki [1] (pp. 131–135) and Koszul [1] for further details.

In the situation where B is nondegenerate, W is a discrete subgroup of the corresponding orthogonal group $G := O(V)$. Now G is a real Lie group, with a Haar measure which provides a notion of volume for the homogeneous space G/W. It can be shown that this volume is finite if and only if B is positive definite (in which case W is finite), or else B has signature $(n-1, 1)$ and $B(\lambda, \lambda) < 0$ for all $\lambda \in C$. When B has signature $(n-1, 1)$, one component of $\{\lambda \in V | B(\lambda, \lambda) = -1\}$ provides a standard model of $(n-1)$-dimensional hyperbolic space.

The finiteness of the volume is expressed by saying that W is a 'lattice' in G. The study of lattices in Lie groups is an old and rich subject, related to the nature of fundamental groups for manifolds on which G acts, and involving such questions as whether the lattice is 'arithmetically' defined. It is useful to distinguish those lattices W (called 'uniform' or 'co-compact') for which G/W is compact. In the hyperbolic case these were classified by Lannér [1] in his thesis. Here there is a very neat criterion, refining the above proposition, which we state without proof:

> W *is compact hyperbolic if and only if both conditions hold:*
>
> (a) B *is nondegenerate, but not positive definite.*
>
> (b) *For each* $s \in S$, *the Coxeter graph obtained by removing* s *from* Γ *is positive definite.*

Remark. Hyperbolic Coxeter groups as defined here provide interesting examples for the study of discrete groups acting on real hyperbolic spaces. But they are not the only Coxeter groups which arise as discrete groups generated by 'reflections' in hyperbolic spaces: they are the particular ones having a simplex as fundamental domain. See the papers of Vinberg for the general theory of hyperbolic reflection groups.

6.9 List of hyperbolic Coxeter groups

Armed with Proposition 6.8 and the remarks following it, one can hope to determine all connected Coxeter graphs for which W is hyperbolic (resp. compact hyperbolic). It is a lengthy exercise to fill in the details rigorously, but the end result (when $n \geq 4$) is summarized below in Figures 2 and 3. By direct inspection, each exhibited graph meets conditions (a) and (b) of the proposition. Less trivial is the verification that no graphs have been overlooked. The list here is taken from Chein [1], who developed an algorithm which was programmed for a computer by N. Spiridon; a version of this list appears in Koszul [1]. (There are some obvious misprints both in Koszul's list and in the exercises in Bourbaki [1]. We invite the skeptical reader to do an independent check!)

The most striking facts about the classification are these: hyperbolic Coxeter groups exist only in ranks 3 to 10, and there are only finitely many in each of ranks 4 to 10. (This is related to the fact that the 'exceptional' types of positive semidefinite Coxeter graphs have rank ≤ 9.) Compact hyperbolic groups exist only in ranks 3, 4, 5.

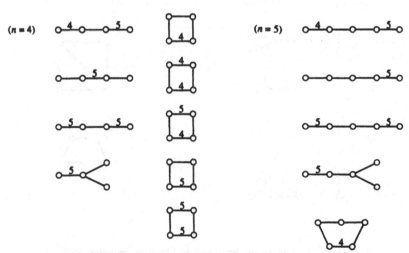

Figure 2: Compact hyperbolic Coxeter groups $(n \geq 4)$

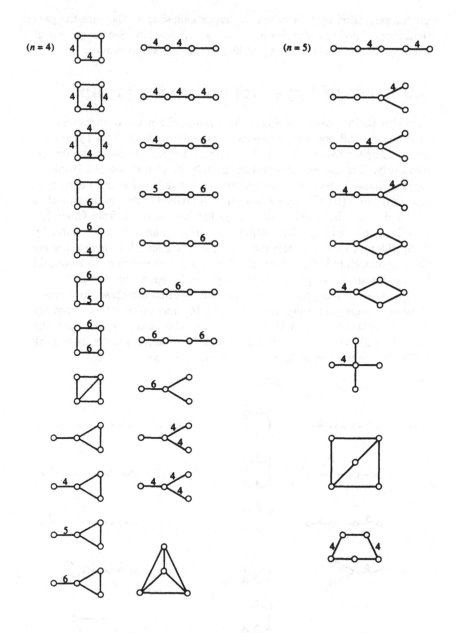

Figure 3: Noncompact hyperbolic Coxeter groups ($n \geq 4$)

($n = 6$)

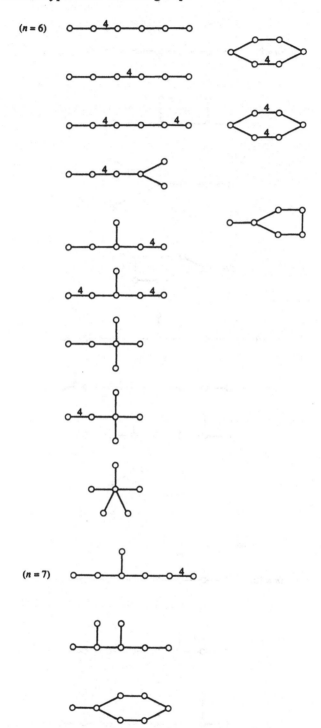

($n = 7$)

($n = 8$)

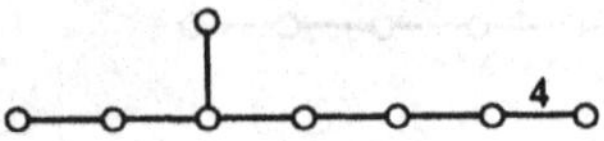

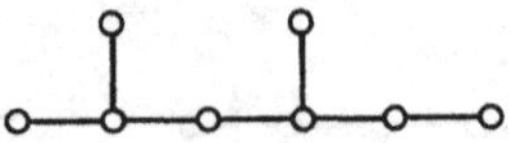

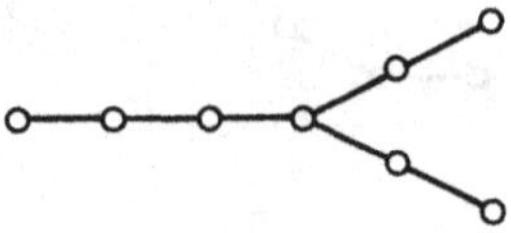

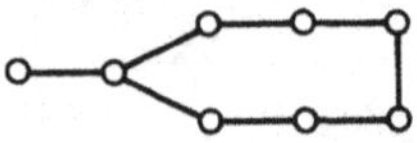

($n = 9$)

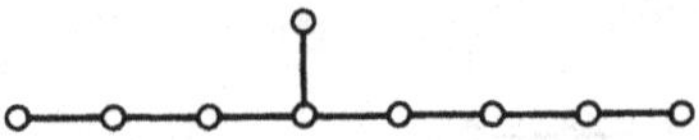

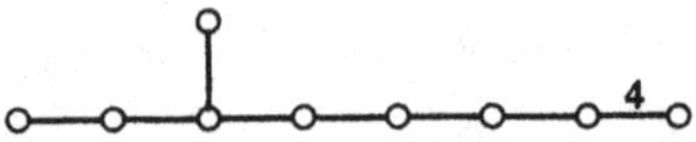

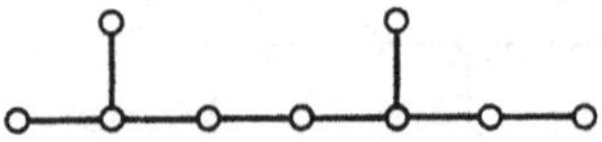

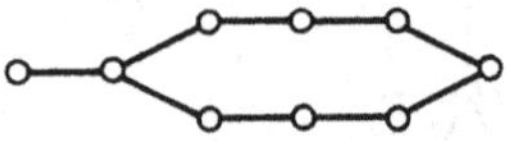

($n = 10$)

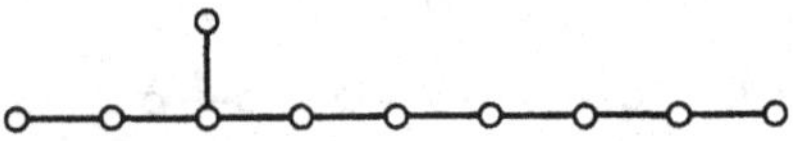

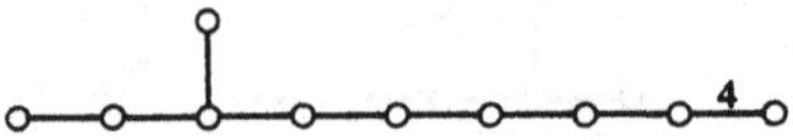

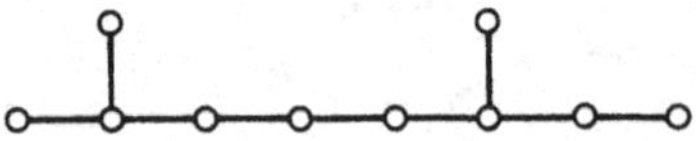

Chapter 7

Hecke algebras and Kazhdan–Lusztig polynomials

This chapter is an introduction to the fundamental paper of Kazhdan–Lusztig [1]. We begin with some generalities about Hecke algebras, which arise in the study by Iwahori [1] and Iwahori–Matsumoto [1] of certain groups of Lie type. The underlying idea is to replace the problem of decomposing an induced representation by the equivalent problem of determining irreducible representations of the associated algebra of intertwining operators (the Hecke algebra); see Curtis [1]–[3]. Here the Hecke algebra is a sort of deformation of the group algebra of the related Weyl group or affine Weyl group. The later work shares this philosophy, but is appreciably more subtle. In any case, what we do is hard to motivate strictly in terms of Coxeter groups.

The treatment in 7.1–7.3 follows Couillens [1] (expanding Bourbaki [1], IV, §2, Exercise 23), but the remainder of the chapter is drawn almost entirely from Kazhdan–Lusztig [1], with added references to work influenced by theirs. Throughout the chapter (W, S) is an arbitrary Coxeter system. Both letters s and t may be used for elements of S, while u, v, w, x, y, z will be used to denote arbitrary elements of W.

7.1 Generic algebras

We begin with a very general construction of associative algebras over a commutative ring A (with 1). Such an algebra will have a free A-basis parametrized by the elements of W, together with a multiplication law which reflects in a certain way the multiplication in W. The algebra will also depend on some parameters $a_s, b_s \in A$ ($s \in S$), subject only to the requirement that $a_s = a_t$ and $b_s = b_t$ whenever s and t are conjugate in W. The starting point for the construction is a free A-module $\mathcal{E}$ on the set W, with basis elements denoted T_w ($w \in W$).

Theorem   *Given elements a_s, b_s as above, there exists a unique structure of associative A-algebra on the free A-module $\mathcal{E}$, with T_1 acting as the identity, such that the following conditions hold for all $s \in S, w \in W$:*

$$T_s T_w = T_{sw} \quad \text{if } \ell(sw) > \ell(w), \tag{1}$$

$$T_s T_w = a_s T_w + b_s T_{sw} \quad \text{if } \ell(sw) < \ell(w). \tag{2}$$

The algebra described by the theorem, denoted $\mathcal{E}_A(a_s, b_s)$, will be called a **generic algebra**. The proof of the theorem will occupy the next two sections. Here we shall make some preliminary remarks and then sketch the idea of the proof.

The group algebra $A[W]$ is one familiar example of a generic algebra: set all $a_s = 0$ and all $b_s = 1$. Another choice of parameters leads to a 'Hecke algebra', to be defined in 7.4 below; this will be the main focus of the present chapter.

Note that the 'right-handed' versions of (1) and (2) will follow from the theorem, using induction on $\ell(w)$. Consider the case $\ell(wt) > \ell(w)$, $t \in S$. Find $s \in S$ for which $\ell(sw) < \ell(w)$. Evidently $\ell(w) = \ell((sw)t) > \ell(sw)$, so induction yields $T_{sw} T_t = T_{swt}$. By (1), $T_s T_{sw} = T_w$, so multiplying both sides by T_t yields:

$$T_w T_t = T_s T_{sw} T_t = T_s T_{swt} = T_{wt},$$

again by (1). Consider the contrary case $\ell(wt) < \ell(w)$. By the right-handed version of (1) just proved, $T_{wt} T_t = T_w$. Multiply both sides by T_t to obtain:

$$T_w T_t = T_{wt} T_t^2 = T_{wt}(a_t T_t + b_t T_1) = a_t T_{wt} T_t + b_t T_{wt} = a_t T_w + b_t T_{wt},$$

using (2) to compute T_t^2.

Later on it will be helpful to have another set of relations equivalent to (1) and (2), as follows:

$$T_s T_w = T_{sw} \quad \text{if } \ell(sw) > \ell(w), \tag{3}$$

$$T_s^2 = a_s T_s + b_s T_1. \tag{4}$$

Obviously (3) and (4) are consequences of (1) and (2). Conversely, assume that $\mathcal{E}$ has an algebra structure (with T_1 as identity element) satisfying (3) and (4). We have to verify (2), in case $\ell(sw) < \ell(w)$. Note that, when $\ell(w) = 1$, we must have $w = s$, so (2) is just (4). In general,

we have $\ell(s(sw)) > \ell(sw)$, so (3) yields $T_s T_{sw} = T_w$. Then use (4) to get
$T_s T_w = T_s^2 T_{sw} = (a_s T_s + b_s T_1) T_{sw} = a_s T_s T_{sw} + b_s T_{sw} = a_s T_w + b_s T_{sw}$,
as required.

Suppose for a moment that $\mathcal{E}$ does admit an algebra structure satisfying (1) and (2). Iteration of (1) shows that $T_w = T_{s_1} \cdots T_{s_r}$ whenever $w = s_1 \cdots s_r$ is a reduced expression. So $\mathcal{E}$ is in fact generated as an algebra by the T_s ($s \in S$), together with $1 = T_1$. In turn, iteration of conditions (1) and (2) enables us to write down the full multiplication table for the basis elements T_w of $\mathcal{E}$. So the *uniqueness* assertion in the theorem is clear.

As to the *existence* of an algebra structure, it is very awkward to introduce directly into the A-module $\mathcal{E}$ the extra structure required. Instead, we exploit the existing ring structure in End $\mathcal{E}$, the algebra of all A-module endomorphisms of $\mathcal{E}$. If $\mathcal{E}$ has an algebra structure, the left multiplication operators corresponding to elements of $\mathcal{E}$ will generate an isomorphic copy of this algebra inside End $\mathcal{E}$. So it will be enough to locate the appropriate subalgebra of End $\mathcal{E}$. To conform with the relations (1) and (2), the left multiplication operators λ_s corresponding to the elements T_s ($s \in S$) would have to behave as follows:

$$\lambda_s(T_w) = T_{sw} \quad \text{if } \ell(sw) > \ell(w), \tag{5}$$
$$\lambda_s(T_w) = a_s T_w + b_s T_{sw} \quad \text{if } \ell(sw) < \ell(w). \tag{6}$$

Similarly, there would have to be right multiplication operators ρ_t ($t \in S$) behaving according to symmetric rules:

$$\rho_t(T_w) = T_{wt} \quad \text{if } \ell(wt) > \ell(w), \tag{7}$$
$$\rho_t(T_w) = a_t T_w + b_t T_{wt} \quad \text{if } \ell(wt) < \ell(w). \tag{8}$$

Now we can sketch briefly how the proof of the theorem goes. Simply define endomorphisms λ_s and ρ_s ($s \in S$) of $\mathcal{E}$ by the preceding rules (extending by linearity from the given action on the basis). Check that every λ_s commutes with every ρ_t ($s, t \in S$), which will be done in 7.2. Then map the subalgebra of End $\mathcal{E}$ (with 1) generated by all λ_s onto $\mathcal{E}$ by sending an endomorphism to its value at T_1. Use the commuting property to see that this map is actually one-to-one, which allows the algebra structure to be transferred to $\mathcal{E}$. Finally, verify the relations (1) and (2).

7.2   Commuting operators

In order to prove that every operator λ_s commutes with every operator ρ_t ($s, t \in S$), we have to sort out a number of configurations of lengths. The following lemma will simplify the task somewhat.

Lemma *Let $w \in W$ and $s, t \in S$. If $\ell(swt) = \ell(w)$ and $\ell(sw) = \ell(wt)$, then $sw = wt$ (or equivalently, $swt = w$).*

Proof. Write $w = s_1 \cdots s_r$ (reduced). There are two possibilities:

(a) $\ell(sw) > \ell(w)$. Then $\ell(w) = \ell((sw)t) < \ell(sw)$, so the Exchange Condition applies to the pair sw, t. We obtain $sw = w't$, where either $w' = ss_1 \cdots \hat{s}_i \cdots s_r$ or else $w' = w$. The first alternative is impossible, since it would imply $w = s(sw) = s(w't) = s_1 \cdots \hat{s}_i \cdots s_r t$, forcing $wt = s_1 \cdots \hat{s}_i \cdots s_r$ to be shorter than w, contrary to the hypothesis $\ell(wt) = \ell(sw) > \ell(w)$. Therefore $w' = w$ and $sw = wt$.

(b) $\ell(sw) < \ell(w) = \ell(s(sw))$. Observe that the hypotheses of the lemma are now satisfied by sw in place of w, so we may apply the result of case (a) to sw. Conclusion: $s(sw) = (sw)t$, i.e., $w = swt$. $\square$

Proposition *For all $s, t \in S$, the operators λ_s and ρ_t commute.*

Proof. Fix $w \in W$ and compare the effects of the two operators $\lambda_s \rho_t$ and $\rho_t \lambda_s$ on T_w. Keeping in mind that multiplication by s or t changes length by 1, we see readily that there are just six possibilities for the relative lengths of sw, wt, swt, w. For example, it is impossible for all of these to have distinct lengths. We analyze each possibility in turn.

(a) $\ell(w) < \ell(wt) = \ell(sw) < \ell(swt)$.

The description of the operators in 7.1 shows at once that $\lambda_s \rho_t(T_w) = T_{swt} = \rho_t \lambda_s(T_w)$.

(b) $\ell(swt) < \ell(wt) = \ell(sw) < \ell(w)$.

By direct calculation, $\lambda_s \rho_t(T_w) = \lambda_s(a_t T_w + b_t T_{wt}) = a_t \lambda_s(T_w) + b_t \lambda_s(T_{wt}) = a_t(a_s T_w + b_s T_{sw}) + b_t(a_s T_{wt} + b_s T_{swt}) = a_t a_s T_w + a_t b_s T_{sw} + b_t a_s T_{wt} + b_t b_s T_{swt}$. An entirely similar calculation of $\rho_t \lambda_s(T_w)$ yields the same result.

(c) $\ell(wt) = \ell(sw) < \ell(swt) = \ell(w)$.

Here we can invoke the above lemma to get $sw = wt$, which says that s is conjugate to t in W, forcing $a_s = a_t$ and $b_s = b_t$. Substitute these into the results of direct calculation: $\lambda_s \rho_t(T_w) = a_t a_s T_w + a_t b_s T_{sw} + b_t T_{swt}$, whereas $\rho_t \lambda_s(T_w) = a_s a_t T_w + a_s b_t T_{wt} + b_s T_{swt}$.

(d) $\ell(wt) < \ell(w) = \ell(swt) < \ell(sw)$.

Here we get $\lambda_s \rho_t(T_w) = a_t T_{sw} + b_t T_{swt} = \rho_t \lambda_s(T_w)$.

(e) $\ell(sw) < \ell(w) = \ell(swt) < \ell(wt)$.

Here $\lambda_s \rho_t(T_w) = a_s T_{wt} + b_s T_{swt} = \rho_t \lambda_s(T_w)$.

(f) $\ell(w) = \ell(swt) < \ell(wt) = \ell(sw)$.

Just as in case (e), we have $\lambda_s \rho_t(T_w) = a_s T_{wt} + b_s T_{swt}$. But $\rho_t \lambda_s(T_w) = a_t T_{sw} + b_t T_{swt}$. We again invoke the above lemma to get $sw = wt$ and $a_s = a_t, b_s = b_t$. $\square$

7.3 Conclusion of the proof

We follow the outline in 7.1. Define $\mathcal{L}$ to be the subalgebra of End $\mathcal{E}$ (with 1) generated by the endomorphisms λ_s ($s \in S$). Then define a map $\varphi : \mathcal{L} \to \mathcal{E}$ by the formula $\varphi(\lambda) = \lambda(T_1)$, thus sending 1 to T_1 and λ_s to T_s for all $s \in S$. It is obvious that φ is an A-module map. Moreover, it is surjective, since all basis elements T_w of $\mathcal{E}$ lie in the image: if $w = s_1 \cdots s_r$ (reduced), then clearly $T_w = \varphi(\lambda_{s_1} \cdots \lambda_{s_r})$.

To show that φ is injective, suppose that $\varphi(\lambda) = 0$, i.e., $\lambda(T_1) = 0$. We use induction on $\ell(w)$ to show that $\lambda(T_w) = 0$ for all $w \in W$, whence $\lambda = 0$. If $\ell(w) > 0$, find $t \in S$ for which $\ell(wt) < \ell(w)$. Thanks to Proposition 7.2, the endomorphism ρ_t commutes with $\mathcal{L}$, in particular with λ. Therefore $\lambda(T_w) = \lambda(T_{(wt)t}) = \lambda(\rho_t(T_{wt})) = \rho_t(\lambda(T_{wt})) = 0$ by induction.

Now that φ is known to be an isomorphism of A-modules, it follows at once that $\mathcal{L}$ has a free A-basis consisting of all $\lambda_w := \lambda_{s_1} \cdots \lambda_{s_r}$ ($w \in W$), where $w = s_1 \cdots s_r$ is reduced and the endomorphism λ_w is independent of the choice of reduced expression. (Here λ_1 is the identity on $\mathcal{E}$.) Moreover, the algebra structure on $\mathcal{L}$ can be transferred to $\mathcal{E}$. It remains only to check that this structure satisfies the relations (1) and (2) in 7.1. The equivalent relations (3) and (4) are actually a bit easier to work with.

Say $\ell(sw) > \ell(w)$. We have to verify that $\lambda_s \lambda_w = \lambda_{sw}$. Taking a reduced expression as above for w, $ss_1 \cdots s_r$ is clearly a reduced expression for the longer element sw. Now $\lambda_s \lambda_w = \lambda_s \lambda_{s_1} \cdots \lambda_{s_r}$ agrees with the definition of λ_{sw}.

The other relation to be verified reads: $\lambda_s^2 = a_s \lambda_s + b_s \lambda_1$. It is enough to check both sides at a typical basis element T_w of $\mathcal{E}$. In case $\ell(sw) > \ell(w)$, we have:

$$\lambda_s^2(T_w) = \lambda_s(T_{sw}) = a_s T_{sw} + b_s T_w = (a_s \lambda_s + b_s \lambda_1)(T_w).$$

In case $\ell(sw) < \ell(w)$, we get instead:

$$\lambda_s^2(T_w) = \lambda_s(a_s T_w + b_s T_{sw}) = a_s \lambda_s(T_w) + b_s T_s T_{sw} = (a_s \lambda_s + b_s \lambda_1)(T_w). \square$$

Exercise 1. Mapping T_w to $T_{w^{-1}}$ defines an A-module automorphism of $\mathcal{E}$. Prove that it is an anti-automorphism of any generic algebra based on $\mathcal{E}$.

Exercise 2. Show that the generic algebra $\mathcal{E}_A(a_s, b_s)$ has a presentation as the A-algebra with 1 generated by the elements T_s ($s \in S$), subject only to the relations: $T_s^2 = a_s T_s + b_s T_1$, $(T_s T_t)^q = (T_t T_s)^q$ if $m(s, t) = 2q < \infty$, and $(T_s T_t)^q T_s = (T_t T_s)^q T_t$ if $m(s, t) = 2q + 1 < \infty$ (as s, t range over S).

7.4   Hecke algebras and inverses

Having constructed generic algebras $\mathcal{E}_A(a_s, b_s)$ over an arbitrary commutative ring A, we now make a special choice: until further notice A will be the ring $\mathbf{Z}[q, q^{-1}]$ of Laurent polynomials over $\mathbf{Z}$ in the indeterminate q. With the further convention that $a_s = q - 1$ and $b_s = q$ for all $s \in S$, we write $\mathcal{H}$ for the resulting generic algebra and call it the **Hecke algebra** of W. (There should be no confusion with the use of $\mathcal{H}$ in Chapter 4 to denote a collection of hyperplanes.) Although we are assigning the same value of a_s (resp. b_s) to all $s \in S$, it should be emphasized that for some applications of Kazhdan–Lusztig theory, more complicated versions are needed (cf. Lusztig [4]). For this reason we allowed more flexibility in the earlier set-up.

The relations (3) and (4) in 7.1 now become:

$$T_s T_w = T_{sw} \text{ if } \ell(sw) > \ell(w),$$

$$T_s^2 = (q - 1)T_s + qT_1.$$

Exercise. Prove that the assignment $T_s \mapsto -1$ ($s \in S$) induces a ring homomorphism $\mathcal{H} \to A$, sending T_w to $(-1)^{\ell(w)}$.

The first special feature to notice in $\mathcal{H}$ is the existence of inverses for the basis elements T_w, because of the presence of q^{-1}. Indeed, the relations imply that for all $s \in S$:

$$T_s^{-1} = q^{-1}T_s - (1 - q^{-1})T_1. \tag{9}$$

If $w = s_1 \cdots s_r$ (reduced expression), we know that $T_w = T_{s_1} \cdots T_{s_r}$. Therefore every T_w is invertible in $\mathcal{H}$. However, as $\ell(w)$ increases it will be progressively more complicated to work out the inverse explicitly as a linear combination of the canonical basis of $\mathcal{H}$. What we can do in this direction introduces an important family of polynomials (the 'R-polynomials').

We shall show that the inverse of a typical $T_{w^{-1}}$ can be written as a combination of those T_x for which $x \leq w$ in the Bruhat ordering (5.9). To do this we need to know how various elements of W are related in the ordering:

Lemma *Let $s \in S, w \in W$ satisfy $sw < w$. Suppose $x < w$.*
(a) *If $sx < x$, then $sx < sw$.*
(b) *If $sx > x$, then $sx \leq w$ and $x \leq sw$.*
Thus, in either case, $sx \leq w$.

Proof. In both cases we rely on Theorem 5.10, which characterizes the Bruhat ordering in terms of subexpressions. Thanks to the Exchange Condition, the hypothesis $sw < w$ implies that w has a reduced expression $w = s_1 \cdots s_r$ with $s_1 = s$. Since x is a subexpression of w, either x is a subexpression of sw or else a subexpression for x begins with s, forcing sx to occur as a subexpression of the (reduced) expression $sw = s_2 \cdots s_r$, whence $sx \leq sw$. In either case, (a) follows. Similarly, in either case, (b) follows. $\square$

To avoid excessive parentheses, we henceforth write $\varepsilon_w = (-1)^{\ell(w)}$ in place of the notation $\varepsilon(w)$ used in Chapter 5. Similarly, we write q_w in place of $q^{\ell(w)}$. When dealing with a polynomial in q alone, we allow ourselves to write F in place of $F(q)$ when no confusion can result.

Proposition *For all $w \in W$,*

$$(T_{w^{-1}})^{-1} = \varepsilon_w q_w^{-1} \sum_{x \leq w} \varepsilon_x R_{x,w}(q) T_x,$$

where $R_{x,w}(q) \in \mathbf{Z}[q]$ is a polynomial of degree $\ell(w) - \ell(x)$ in q, and where $R_{w,w}(q) = 1$.

Proof. This is clear when $w = 1$. Thanks to (9), it is also clear when $w = s$ lies in S, if we set $R_{1,s} := q - 1$. Proceed by induction on $\ell(w)$. In the course of the proof, we shall actually obtain an algorithm for computing the R-polynomials, about which more will be said in 7.5. For convenience, define $R_{x,w}$ to be 0 whenever $x \not\leq w$.

Assuming $\ell(w) > 0$, we can write $w = sv$ for some $s \in S$, with $\ell(v) < \ell(w)$. Thus $\varepsilon_w = -\varepsilon_v$ and $q_w = q_v q$. Using the induction hypothesis, we compute:

$$
\begin{aligned}
(T_{w^{-1}})^{-1} &= (T_{v^{-1}} T_s)^{-1} \\
&= T_s^{-1} (T_{v^{-1}})^{-1} \\
&= q^{-1}(T_s - (q - 1)T_1)(\varepsilon_v q_v^{-1} \sum_{y \leq v} \varepsilon_y R_{y,v} T_y) \\
&= \varepsilon_w q_w^{-1} \left[(q - 1) \sum_{y \leq v} \varepsilon_y R_{y,v} T_y - \sum_{y \leq v} \varepsilon_y R_{y,v} T_s T_y \right] .(10)
\end{aligned}
$$

The second sum in (10) involves two sorts of terms. If $sy > y$, we just get $\varepsilon_y R_{y,v} T_{sy}$. But if $sy < y$, we get instead:

$$(q - 1)\varepsilon_y R_{y,v} T_y + q\varepsilon_y R_{y,v} T_{sy},$$

the first term of which cancels a term in the first sum in (10). This allows us to rewrite (10) as a sum over three kinds of terms:

$$y \leq v, \ y < sy \qquad (q-1)\varepsilon_y R_{y,v} T_y \qquad\qquad (11)$$

$$y \leq v, \ y < sy \qquad -\varepsilon_y R_{y,v} T_{sy} \qquad\qquad (12)$$

$$y \leq v, \ y > sy \qquad -q\varepsilon_y R_{y,v} T_{sy} \qquad\qquad (13)$$

In each case, we have $y < w$ and (thanks to the lemma above with y in the role of x) $sy \leq w$. Notice too that every $x \leq w$ occurs either as a $y \leq v$ or as an sy with $y \leq v$ (or both), because of 5.10 again. So it remains only to check that the coefficient of T_x in (10) meets the criterion of the proposition.

Consider $x \leq w$ with $x > sx$. Then T_x occurs only in case (12), with $x = sy$ for $y \leq v$, and we obtain as coefficient: $-\varepsilon_y R_{y,v} = \varepsilon_x R_{sx,sw}$, of degree $\ell(sw) - \ell(sx) = \ell(w) - \ell(x)$. In the extreme case $x = w$, note that $y = v$ and $R_{v,v} = 1$ by induction. So the polynomial $R_{x,w} := R_{sx,sw}$ has the required features.

Consider the contrary case $x < w$, with $x < sx$. Here there are two possibilities:

(a) In case $sx < v$, T_x occurs both in a term of type (11), with $x = y \leq v$, and in a term of type (13), with $x = sy, y = sx \leq v$. The combined coefficient is

$$(q-1)\varepsilon_x R_{x,v} - q\varepsilon_{sx} R_{sx,v}.$$

However note that deg $qR_{sx,v} = \ell(v) - \ell(sx) + 1 = (\ell(w) - 1) - (\ell(x)+1)+1 = \ell(w)-\ell(x)-1$, whereas deg $(q-1)R_{x,v} = \ell(v)-\ell(x)+1 = (\ell(w) - 1) - \ell(x) + 1 = \ell(w) - \ell(x)$. Thus the combined coefficient has degree $\ell(w) - \ell(x)$ as required, and we can define

$$R_{x,w} := (q-1)R_{x,sw} + qR_{sx,sw}.$$

(b) In case $sx \not\leq v$, T_x occurs only in a term of type (11), with coefficient equal to $\varepsilon_x(q-1)R_{x,v}$. Using the convention $R_{sx,v} = 0$, we see that $R_{x,w}$ can be defined precisely as in case (a). This completes the induction step. □

It is worth remarking that the Bruhat ordering is forced on us by the way inversion works in $\mathcal{H}$, whether or not we have previously felt motivated to introduce this ordering for arbitrary Coxeter groups. Indeed, the proposition shows that $R_{x,w}$ is nonzero if and only if $x \leq w$.

7.5   Computing the R-polynomials

Let us first make explicit the algorithm for computing $R_{x,w}$ implied by the proof of Proposition 7.4. The idea is to use induction on $\ell(w)$, starting with the fact that $R_{w,w} = 1$ for all $w \in W$, while $R_{x,w} = 0$ unless $x \leq w$. For the induction step, we need to compute $R_{x,w}$, assuming that all polynomials $R_{y,z}$ are known for $\ell(z) < \ell(w)$. Fix $s \in S$ for which $sw < w$. Then two configurations have to be dealt with, as in Lemma 7.4:

(A) $x < w, sx < x$ (forcing $sx < sw$). Here we found that $R_{x,w} = R_{sx,sw}$, which is already known since $sw < w$.

(B) $x < w, x < sx$ (forcing $sx \leq w$ and $x \leq sw$). Here we found that $R_{x,w} = (q-1)R_{x,sw} + qR_{sx,sw}$, both terms of which are already known. (Recall that the first term has degree $\ell(w) - \ell(x)$, while the second term has lower degree and might be 0.)

It is sometimes useful to have alternate versions of (A) and (B), with s occurring on the right rather than the left. These follow from a 'symmetric' version of the proof of Proposition 7.4, as the reader can check. For example:

(A') $x < w, xs < x, ws < w$ (forcing $xs < ws$). Then $R_{x,w} = R_{xs,ws}$.

Exercise 1. Prove that $R_{x,w}(1) = 0$ unless $x = w$.

To get a better feeling for the way the R-polynomials are built up inductively, we consider the special case $\ell(w) - \ell(x) = 1$. If $w = s_1 \cdots s_r$ is a reduced expression, we can obtain x by omitting a single s_i. Repeated use of (A) and (A') reduces matters to the situation: $w = s_i, x = 1$. As remarked at the beginning of the proof of Proposition 7.4, we get $R_{x,w} = q - 1$ from the explicit formula (9) for T_s^{-1}.

To carry this a step further, consider what happens when $\ell(w) - \ell(x) = 2$. Fixing as before a reduced expression for w, we observe that (for reasons of parity) x can be obtained by omitting precisely two of the factors s_i, s_j ($i < j$). Again we can apply (A) and (A') repeatedly to reduce to the case: $w = s_i \cdots s_j, x = s_{i+1} \cdots s_{j-1}$. Taking $s = s_i$, we have the configuration: $sw < w, sx > x$. Therefore (B) applies and we have $R_{x,w} = (q-1)R_{x,sw} + qR_{sx,sw}$. The first term is known from the preceding calculation: $R_{x,sw} = q - 1$. On the other hand, both sx and sw have the same length but are unequal, forcing the second term to be 0. Conclusion: $R_{x,w} = (q-1)^2$.

The intrepid reader may wish to press on with these explicit calculations. However, they rapidly become less manageable, because of the more complicated possibilities for subexpressions when more than two factors are omitted. For example, if $\ell(w) - \ell(x) = 3$, it is possible that x is obtained by omitting only one factor in a reduced expression for w. We shall be content to quote here a result of Deodhar [5], which gives a closed formula for the R-polynomials (still involving some intricate calculations with subexpressions). The formula expresses $R_{x,w}$ as a sum of terms of the form $(q-1)^n q^m$. But the summation is over a set which is defined in a delicate way, as follows.

Fix once and for all a reduced expression $w = s_1 \cdots s_r$, so every $x \leq w$ occurs as a subexpression. Deodhar reformulates the notion of 'subexpression' to mean an element $\sigma = (\sigma_0, \sigma_1, \ldots, \sigma_r) \in W^{r+1}$, where $\sigma_0 = 1, \sigma_j = \sigma_{j-1}$ or $\sigma_{j-1}s_j$ ($1 \leq j \leq r$). Here σ corresponds to the subexpression $s_1 \cdots \widehat{s_{i_1}} \cdots \widehat{s_{i_p}} \cdots s_r$, with $\{i_1, \ldots, i_p\} = \{j | \sigma_j = \sigma_{j-1}\}$; the resulting $x \leq w$ is just σ_r. Now define $n(\sigma) := p$ and define $m(\sigma) := \mathrm{Card} \{j | \sigma_{j-1} > \sigma_j\}$. Finally, consider the set of 'distinguished' subexpressions $D(x)$ consisting of those σ for which $x = \sigma_r$ and $\sigma_j \leq \sigma_{j-1}s_j$ for all $1 \leq j \leq r$. Note that it is not immediately clear from the definitions that $D(x)$ need be nonempty for $x \leq w$. However, this is seen in the proof of Deodhar's formula:

$$R_{x,w} = \sum_{\sigma \in D(x)} (q-1)^{n(\sigma)} q^{m(\sigma)}.$$

Exercise 2. Check that Deodhar's formula agrees with our preceding calculations of $R_{x,w}$ in case $\ell(w) - \ell(x) \leq 2$.

Exercise 3. When $W = S_3$, work out all $R_{x,w}$. Compare the results with Deodhar's formula. For example, when $x = 1, w = w_\circ$ (the longest element), $R_{x,w} = (q-1)^3 + q(q-1)$.

7.6   Special case: finite Coxeter groups

When W is finite, the R-polynomials exhibit some extra symmetry, which will be exploited in 7.13 below. Recall from 1.8 the longest element $w_\circ$ of W, which satisfies $\ell(w_\circ w) = \ell(w_\circ) - \ell(w)$ for all $w \in W$. Recall also from Example 3 of 5.9 that $x < w$ if and only if $w_\circ w < w_\circ x$.

Proposition *If W is finite, then $R_{x,w} = R_{w_\circ w, w_\circ x}$ for all $x \leq w$.*

Proof. Proceed by induction on $\ell(w)$, starting with the fact that $R_{1,1} = R_{w_\circ, w_\circ} = 1$. If $\ell(w) > 0$, find $s \in S$ for which $ws < w$. There are two cases to consider, following the rules in 7.5.

(a) If $xs < x$, then also $w_\circ x < w_\circ xs$, so we can apply (A') to each of these situations and appeal to the induction hypothesis to get:

$$R_{x,w} = R_{xs,ws} = R_{w_\circ ws, w_\circ xs} = R_{w_\circ w, w_\circ x}.$$

(b) If $x < xs$, the situation is more complicated. Using the symmetric version (B') of (B), we have

$$R_{x,w} = (q-1)R_{x,ws} + qR_{xs,ws}.$$

By induction, this is equal to

$$(q-1)R_{w_ows,w_ox} + qR_{w_ows,w_oxs}.$$

Notice that (A′) applies to the first term, allowing us to replace it with $(q-1)R_{w_ow,w_oxs}$. In turn, (B′) applies, yielding R_{w_ow,w_ox}, as required.
□

7.7   An involution on $\mathcal{H}$

Having seen how to invert the basis elements T_w of $\mathcal{H}$, we go a step further by introducing an involution $\iota : \mathcal{H} \to \mathcal{H}$ (a ring automorphism of order 2). To define ι, begin with the involution ι of $A = \mathbf{Z}[q, q^{-1}]$ sending q to q^{-1}, and then define $\iota(T_w) := (T_{w^{-1}})^{-1}$. Combine these assignments and extend additively to obtain a 'semilinear' map $\iota : \mathcal{H} \to \mathcal{H}$.

Using (9) in 7.4, the reader can easily check that $\iota^2(T_s) = T_s$ for all $s \in S$. To prove that ι^2 is the identity on $\mathcal{H}$, it will therefore be enough to prove that ι is a ring homomorphism (since the T_s generate $\mathcal{H}$ as an A-algebra).

First we show:

$$\iota(T_sT_w) = \iota(T_s)\iota(T_w) \quad \text{whenever } s \in S, w \in W. \tag{14}$$

There are two cases to check, the first being straightforward. If $\ell(sw) > \ell(w)$, we just compute directly:

$$\iota(T_sT_w) = \iota(T_{sw}) = (T_{w^{-1}s})^{-1} = (T_{w^{-1}}T_s)^{-1} = T_s^{-1}(T_{w^{-1}})^{-1} = \iota(T_s)\iota(T_w).$$

The second case is more complicated. If $\ell(sw) < \ell(w)$, write $v = (sw)^{-1}$, so $w^{-1} = vs$. Now

$$\iota(T_sT_w) = \iota(qT_{sw} + (q-1)T_w) = q^{-1}T_v^{-1} + (q^{-1}-1)(T_{w^{-1}})^{-1}.$$

Note here that $q^{-1} - 1 = -q^{-1}(q-1)$. Also, $T_{w^{-1}} = T_{vs} = T_vT_s$ has inverse equal to $T_s^{-1}T_v^{-1}$. Further, recall that $T_s^{-1} = q^{-1}(T_s - (q-1)T_1)$. Substituting all of these expressions, we obtain:

$$\iota(T_sT_w) = q^{-2}(q^2 - q + 1)T_v^{-1} - (q-1)q^{-2}T_sT_v^{-1}.$$

On the other hand,

$$\iota(T_s)\iota(T_w) = T_s^{-1}(T_{w^{-1}})^{-1} = (T_s)^{-2}T_v^{-1}.$$

Again we can substitute the expression for T_s^{-1}, and then substitute $T_s^2 = (q-1)T_s + qT_1$, to get the same end result.

Having verified (14), it is easy to conclude by induction on $\ell(w')$ that

$$\iota(T_{w'}T_w) = \iota(T_{w'})\iota(T_w) \quad \text{for all } w, w' \in W.$$

If $\ell(w') > 1$, find $s \in S$ for which $\ell(w's) < \ell(w')$. Then $\iota(T_{w'}T_w) = \iota(T_{w's}T_sT_w) = \iota(T_{w's})\iota(T_sT_w)$ by induction (thinking of T_sT_w as some A-linear combination of basis elements T_v). In turn, the special case (14) together with another application of the induction hypothesis shows that the last expression equals $\iota(T_{w's})\iota(T_s)\iota(T_w) = \iota(T_{w's}T_s)\iota(T_w) = \iota(T_{w'})\iota(T_w)$.

Exercise. Define another involution σ of $\mathcal{H}$, as follows. On A let $\sigma = \iota$, while $\sigma(T_w) := \varepsilon_w q_w^{-1} T_w$. (Imitate the above steps.) Show that $\sigma\iota = \iota\sigma$.

7.8 Further properties of R-polynomials

The involution ι will play a key role in the definition of Kazhdan–Lusztig polynomials in 7.9. To get more familiar with its properties, we develop some further useful facts about the R-polynomials, the first of which requires only the involution on A, and the third of which doesn't mention ι explicitly (but would be harder to prove without using it). To avoid excessive parentheses, we use a bar in place of ι (applied to elements of A) whenever convenient; thus $\bar{R}_{x,w}$ means the same thing as $R_{x,w}(q^{-1})$.

Proposition *For all $x, w \in W$, we have:*
   (a) $\bar{R}_{x,w} = \varepsilon_x \varepsilon_w q_x q_w^{-1} R_{x,w}$.
   (b) $(T_{w^{-1}})^{-1} = \sum_{x \leq w} q_x^{-1} \bar{R}_{x,w} T_x$.
   (c) $\sum_{x \leq y \leq w} \varepsilon_x \varepsilon_y R_{x,y} R_{y,w} = \delta_{x,w}$ (*Kronecker delta*).

Proof. (a) This is checked inductively, following the rules at the beginning of 7.5. In the first case there, we have $x < w, sx < x, sw < w$, so that $R_{x,w} = R_{sx,sw}$ (and similarly after applying ι). By induction,

$$\bar{R}_{sx,sw} = \varepsilon_{sx}\varepsilon_{sw}q_{sx}q_{sw}^{-1}R_{sx,sw} = (-\varepsilon_x)(-\varepsilon_w)q_x q^{-1} q_w^{-1} q R_{x,w}.$$

This yields the desired expression for $\bar{R}_{x,w}$. In the second case, the argument is less direct. Here we have $x < w, x < sx, sw < w$, with

$$R_{x,w} = (q-1)R_{x,sw} + qR_{sx,sw}.$$

Applying ι yields

$$\bar{R}_{x,w} = -q^{-1}(q-1)\bar{R}_{x,sw} + q^{-1}\bar{R}_{sx,sw}.$$

By induction

$$\bar{R}_{x,sw} = \varepsilon_x\varepsilon_{sw}q_x q_{sw}^{-1}R_{x,sw},$$

while

$$\bar{R}_{sx,sw} = \varepsilon_{sx}\varepsilon_{sw}q_{sx}q_{sw}^{-1}R_{sx,sw}.$$

It remains to compare coefficients, using the fact that $q_{sx} = q_x q$ and $q_{sw} = q_w q^{-1}$. The coefficient of $R_{x,sw}$ is

$$-q^{-1}(q-1)\varepsilon_x(-\varepsilon_w)q_x q_w^{-1}q = (q-1)\varepsilon_x\varepsilon_w q_x q_w^{-1},$$

as required. Similarly, the coefficient of $R_{sx,sw}$ is

$$q^{-1}(-\varepsilon_x)(-\varepsilon_w)(q_x q)(q_w^{-1}q) = q\varepsilon_x\varepsilon_w q_x q_w^{-1},$$

as required.

(b) By Proposition 7.4,

$$(T_{w^{-1}})^{-1} = \varepsilon_w q_w^{-1} \sum_{x \leq w} \varepsilon_x R_{x,w} T_x.$$

Now substitute the result of part (a).

(c) Apply ι to the formula in (b), using the variable y in place of x:

$$T_w = \sum_{y \leq w} q_y R_{y,w}(T_{y^{-1}})^{-1}.$$

Then substitute for $(T_{y^{-1}})^{-1}$ the expression in Proposition 7.4 to get:

$$T_w = \sum_{y \leq w} q_y R_{y,w} \varepsilon_y q_y^{-1} \sum_{x \leq y} \varepsilon_x R_{x,y} T_x. \tag{15}$$

Now compare the coefficient of T_x (equal to 0 or 1) on each side of (15). $\square$

Part (c) of the proposition shows how to 'invert' the matrix of R-polynomials. This matrix is infinite if W is; but it has only finitely many nonzero entries in each column and (when written in a way compatible with the Bruhat ordering) it is upper triangular unipotent. In the literature (such as Kazhdan–Lusztig [1]), one may see (c) written with ε_w in place of ε_x; this does not change the formula!

Exercise. Verify part (c) of the proposition directly when $W = S_3$, by using the explicit matrix of R-polynomials.

7.9 Kazhdan–Lusztig polynomials

We now look for a new basis $\{C_w\}$ of the A-module $\mathcal{H}$, indexed again by W, but consisting of elements fixed by the involution ι. We can see how to get started by experimenting with formula (9) in 7.4:

$$T_s^{-1} = q^{-1}T_s - (1 - q^{-1})T_1.$$

It is easy to check that ι sends $T_s - qT_1$ to $q^{-1}(T_s - qT_1)$. If we are willing to introduce a square root of q (written $q^{\frac{1}{2}}$), we therefore have an element fixed by ι for each $s \in S$:

$$C_s := q^{-\frac{1}{2}}(T_s - qT_1). \tag{16}$$

Formally, we replace $\mathbf{Z}[q, q^{-1}]$ by the ring $\mathbf{Z}[q^{\frac{1}{2}}, q^{-\frac{1}{2}}]$ of Laurent polynomials in the indeterminate $q^{\frac{1}{2}}$, so that the previous ring A becomes a subring of the new one. This has no effect on the previous formal calculations in $\mathcal{H}$. *For the remainder of this chapter A denotes the enlarged ring.*

It is tempting to construct further ι-invariants simply by multiplying various $C_s (s \in S)$, in the spirit of the way the original basis elements T_w of $\mathcal{H}$ are built out of the T_s. For example, if $s \neq t$, we find

$$C_s C_t = q^{-1}(T_{st} - qT_s - qT_t + q^2 T_1). \tag{17}$$

We might label this element C_{st}. Note that if $st = ts$ (the only other possible reduced expression!), we would have obtained the same element by multiplying C_t times C_s. But our naive approach becomes less satisfactory if we go a step further (assuming $\ell(sts) = 3$):

$$C_s C_t C_s = q^{-\frac{3}{2}}(T_{sts} - qT_{st} - qT_{ts} + q^2(1+q^{-1})T_s + q^2 T_t - q^3(1+q^{-1})T_1). \tag{18}$$

If we want to label this element C_{sts}, there is an obvious ambiguity in case $sts = tst$, because $C_t C_s C_t \neq C_s C_t C_s$. Moreover, the polynomials appearing as coefficients of T_s and T_1 in (18) are more complicated than we would like. Notice that $C_s C_t C_s - C_s$ is an ι-invariant which is also a linear combination of the T_x for $x \leq sts$; but its coefficients are quite simple. (And in case $sts = tst$, s and t are interchangeable throughout.)

What we are seeking in general is an ι-invariant element C_w which is a linear combination of the T_x for $x \leq w$ (the coefficient of T_w being nonzero) and whose polynomial coefficients are as uncomplicated as possible. The following basic theorem of Kazhdan–Lusztig [1] provides an optimal choice:

Theorem *For each $w \in W$ there exists a unique element $C_w \in \mathcal{H}$ having the following two properties:*

(a) $\iota(C_w) = C_w$,

(b) $C_w = \varepsilon_w q_w^{\frac{1}{2}} \sum \varepsilon_x q_x^{-1} \bar{P}_{x,w} T_x$ *(sum over $x \leq w$), where $P_{w,w} = 1$ and $P_{x,w}(q) \in \mathbf{Z}[q]$ has degree $\leq \frac{1}{2}(\ell(w) - \ell(x) - 1)$ if $x < w$.*

The proof of the theorem will occupy the next two sections. In the meantime, we make some observations about the formulation.

First, the elements $C_1 := T_1$ and $C_s (s \in S), C_{st} (s \neq t)$ defined by (16) and (17) clearly meet requirements (a) and (b). (To check uniqueness directly is a useful exercise at this point.) For $W = S_3$, one can use the discussion after (18) to complete the verification of the theorem; in this case, all $P_{x,w} = 1$.

The polynomials $P_{x,w}$ (which are by no means always equal to 1) turn out to be of fundamental interest. They are called the **Kazhdan–Lusztig polynomials**. They are appreciably more subtle than the earlier $R_{x,w}$. For example, their precise degrees are not readily predictable. It is conjectured in Kazhdan–Lusztig [1] that all coefficients of $P_{x,w}$ are nonnegative, but this remains unproved (at the time of writing) except in some important special cases; see 7.12 below.

As in the case of the R-polynomials, it is convenient to make the convention that $P_{x,w} = 0$ whenever $x \not\leq w$.

Exercise. Assuming the truth of the theorem, prove that the elements C_w ($w \in W$) form a basis of the A-module $\mathcal{H}$.

Remark. The theorem can be reformulated in a way which may look simpler. In place of the elements C_w we can require elements C'_w satisfying (a) together with

$$C'_w = q_w^{-\frac{1}{2}} \sum_{x \leq w} P_{x,w} T_x,$$

with $P_{x,w}$ as in (b). The reader can easily verify that the two formulations are equivalent, by first checking that $C'_w = \varepsilon_w \sigma(C_w)$ (with σ the involution of $\mathcal{H}$ described in the exercise in 7.7). To be consistent with most of the literature, we shall stick with the elements C_w.

7.10 Uniqueness

We first demonstrate for each $w \in W$ the *uniqueness* of the element

$$C_w = \sum_{x \leq w} a(x, w) \bar{P}_{x,w} T_x, \quad \text{where } a(x, w) := \varepsilon_w \varepsilon_x q_w^{\frac{1}{2}} q_x^{-1}, \qquad (19)$$

assuming that C_w has properties (a) and (b) in Theorem 7.9. This amounts to showing that the polynomials $P_{x,w}$ can be chosen in at most one way. For fixed w we proceed by induction on $\ell(w) - \ell(x)$, starting with the requirement that $P_{w,w} = 1$. Thus we may assume that all $P_{y,w}$ are uniquely determined, for $x < y \leq w$. We must show that this forces the choice of $P_{x,w}$.

Start with the formula (19) for C_w, written with y in place of x. Apply ι, replacing q by q^{-1} and T_y by $(T_{y^{-1}})^{-1}$, then substitute for the latter the expression in Proposition 7.4 (with y in place of w there). This yields (after obvious cancellations) a sum over all pairs x, y satisfying $x \leq y \leq w$:

$$\varepsilon_w q_w^{-\frac{1}{2}} \sum \varepsilon_x R_{x,y} P_{y,w} T_x.$$

Next equate the coefficient of a fixed T_x with the original coefficient in C_w to get:

$$\varepsilon_w q_w^{\frac{1}{2}} \varepsilon_x q_x^{-1} \bar{P}_{x,w} = \varepsilon_w q_w^{-\frac{1}{2}} \sum_{x \leq y \leq w} \varepsilon_x R_{x,y} P_{y,w}.$$

Further cancellation of signs and multiplication of both sides by $q_x^{\frac{1}{2}}$ yields:

$$q_w^{\frac{1}{2}} q_x^{-\frac{1}{2}} \bar{P}_{x,w} = q_w^{-\frac{1}{2}} q_x^{\frac{1}{2}} \sum_{x \leq y \leq w} R_{x,y} P_{y,w}. \tag{20}$$

Finally, move the term for $y = x$ to the left (using the fact that $R_{x,x} = 1$):

$$q_w^{\frac{1}{2}} q_x^{-\frac{1}{2}} \bar{P}_{x,w} - q_w^{-\frac{1}{2}} q_x^{\frac{1}{2}} P_{x,w} = q_w^{-\frac{1}{2}} q_x^{\frac{1}{2}} \sum_{x < y \leq w} R_{x,y} P_{y,w}. \tag{21}$$

Assuming that all $P_{y,w}$ $(x < y \leq w)$ are already uniquely determined, we have to argue that $P_{x,w}$ is also determined. Since $x < w$, the degree assumption in (b) implies that the first term on the left is a polynomial in $q^{\frac{1}{2}}$ without constant term, while the second term is a polynomial in $q^{-\frac{1}{2}}$ without constant term. Thus no cancellation occurs, and there is at most one choice for $P_{x,w}$ satisfying (21).

Remark. Note that (21) would provide an algorithm for computation of the Kazhdan–Lusztig polynomials. (The reader might follow through on this for $W = S_3$, to get a better feeling for what it involves.) We shall have more to say in 7.12 about the problem of computing the polynomials efficiently.

Exercise. If $\ell(w) - \ell(x) = 1$, deduce from (21) that $P_{x,w} = 1$ (using the fact that $R_{x,w} = q - 1$ in this case).

7.11   Existence

Now we tackle the *existence* of the elements C_w, which is less straightforward. As we observed in 7.9, multiplication of various elements C_s gives at least a first approximation to what we want, but some 'correction' terms may be needed. The choice of these turns out to be delicate. It depends on which Kazhdan–Lusztig polynomials $P_{x,w}(x < w)$ have the largest allowable degree $\frac{1}{2}(\ell(w) - \ell(x) - 1)$. When this degree is attained (which is of course possible only if $\varepsilon_w = -\varepsilon_x$), we write $x \prec w$ and let $\mu(x, w)$ be the coefficient of the highest power of q in $P_{x,w}$.

It is natural to proceed by induction on $\ell(w)$ in proving the existence of the desired elements C_w. (We have already disposed directly of the cases $\ell(w) \le 2$.) For the induction step, find $s \in S$ for which $\ell(sw) < \ell(w)$ and set $v = sw$. So C_v has already been constructed. Note that:

$$a(x, w) = -q^{\frac{1}{2}} a(x, v).$$

We now define:

$$C_w := C_s C_v - \sum \mu(z, v) C_z, \tag{22}$$

where the sum is taken over only those z which satisfy both $z \prec v$ and $sz < z$. Obviously C_w is ι-invariant. Recall that

$$C_s = q^{-\frac{1}{2}} T_s - q^{\frac{1}{2}} T_1.$$

This makes it clear that C_w is an A-linear combination of elements T_x, $x \le w$. We have to look closely at the coefficient of T_x for each fixed x, starting with $x = w$. Evidently T_w occurs only in the product $T_s C_v$, with coefficient $q^{-\frac{1}{2}} a(v, v) \bar{P}_{v,v} = q^{-\frac{1}{2}} q_v^{\frac{1}{2}} q_v^{-1} = q_w^{-\frac{1}{2}}$. This agrees with (19) when $x = w$ (with $P_{w,w} = 1$).

Next fix $x < w$. T_x might occur in two ways in $C_s C_v$, either straightforwardly in C_v (if $x \le v$) or else indirectly in $T_s C_v$ when T_s is multiplied by T_{sx} (if $sx \le v$). We distinguish two cases.

First suppose $x < sx$, so that $T_s T_{sx} = q T_x + (q-1) T_{sx}$ and $q^{-\frac{1}{2}} T_s C_v$ involves T_x with the coefficient:

$$q^{-\frac{1}{2}} q a(sx, v) \bar{P}_{sx,v} = q^{\frac{1}{2}}(-1^{-1}) a(x, v) \bar{P}_{sx,v} = q^{-1} a(x, w) \bar{P}_{sx,v}.$$

On the other hand, $-q^{\frac{1}{2}} T_1 C_v$ involves T_x with the coefficient:

$$-q^{\frac{1}{2}} a(x, v) \bar{P}_{x,v} = a(x, w) \bar{P}_{x,v}.$$

Combining these, we see that the coefficient of T_x in $C_s C_v$ is

$$q^{-1} a(x, w) \bar{P}_{sx,v} + a(x, w) \bar{P}_{x,v}.$$

Instead suppose $sx < x$, so $T_s T_{sx} = T_x$ and $T_s T_x = q T_{sx} + (q-1) T_x$. In $q^{-\frac{1}{2}} T_s C_v$ we therefore get respective coefficients of T_x equal to:

$$q^{-\frac{1}{2}} a(sx, v) \bar{P}_{sx,v} = q^{-\frac{1}{2}}(-q) a(x, v) \bar{P}_{sx,v} = a(x, w) \bar{P}_{sx,v},$$

$$(q-1) q^{-\frac{1}{2}} a(x, v) \bar{P}_{x,v} = (q^{-1} - 1) a(x, w) \bar{P}_{x,v}.$$

On the other hand, $-q^{\frac{1}{2}} T_1 C_v$ involves T_x with the coefficient:

$$-q^{\frac{1}{2}} a(x, v) \bar{P}_{x,v} = a(x, w) \bar{P}_{x,v}.$$

Combining these, we see that the coefficient of T_x in $C_s C_v$ is

$$a(x, w) \bar{P}_{sx,v} + q^{-1} a(x, w) \bar{P}_{x,v}.$$

Finally, the coefficient of T_x in $-\sum \mu(z,v)C_v$ is always of the form

$$-\sum \mu(z,v)a(x,z)\bar{P}_{x,z} = -\sum \mu(z,v)q_z^{\frac{1}{2}}q_w^{-\frac{1}{2}}a(x,w)\bar{P}_{x,z},$$

using the fact that $\varepsilon_z\varepsilon_w = 1$ when $z \prec v = sw$.

If we set $c = 0$ when $x < sx$ and $c = 1$ when $sx < x$, we can combine these calculations to express C_w in the form (19), with

$$P_{x,w} := q^{1-c}P_{sx,v} + q^c P_{x,v} - \sum \mu(z,v)q_z^{-\frac{1}{2}}q_w^{\frac{1}{2}}P_{x,z}, \qquad (23)$$

where as in (22) the summation is over those $z \prec v$ for which $sz < z$ (with the convention that $P_{x,z} = 0$ unless $x \le z$).

Now a careful scrutiny of the terms in (23), using the inductive information about the previously defined polynomials, will show that $P_{x,w}$ has degree at most $\frac{1}{2}(\ell(w) - \ell(x) - 1)$. This is mostly routine to check, except for the case $sx < x, c = 1$, when the middle term $qP_{x,v}$ could have degree exactly $1 + \frac{1}{2}(\ell(v) - \ell(x) - 1) = \frac{1}{2}(\ell(w) - \ell(x))$, which is too large. But in this case we have $x \prec v$, and (since $sx < x$) there is a term for $z = x$ in the sum in (23) which is precisely equal to the highest degree term of $qP_{x,v}$ (thanks to $P_{x,x} = 1$). So this cancels the offending term. Moreover, only in this situation does the sum involve $P_{x,x}$, so all terms in the sum have correctly bounded degrees. This completes the proof of the theorem. $\square$

Remark. We have reproduced here the original proof in Kazhdan–Lusztig [1], because it yields very explicit information about the polynomials $P_{x,w}$ and the way in which the elements C_w multiply. However, Lusztig [4] sketches a more elegant existence proof (suggested by O. Gabber), as follows. The idea is to reverse the steps in the uniqueness proof above, by showing inductively that (21) can be solved for $P_{x,w}$. (Then the bound on its degree will follow, and one can define an ι-invariant C_w as in (19).)

The problem is to show that applying ι to the right side of (21) just changes the sign (since this must be true of the left hand side). So apply ι directly, then substitute for $\bar{P}_{y,w}$ the formula of type (20) already known inductively, and finally use the inversion formula for R-polynomials (part (c) of Proposition 7.8).

Exercise. Use (23) to show that $P_{x,w}(0) = 1$ for all $x \le w$. If $\ell(w) - \ell(x) \le 2$, deduce that $P_{x,w} = 1$.

7.12   Examples

In principle one can use (21) in 7.10 or (23) in 7.11 to compute Kazhdan–Lusztig polynomials recursively (keeping in mind Lemma 7.4). But only in the simplest cases can one get definitive results in this way without the help of a computer. Here are some indications of what has been done.

(a) Suppose W is a dihedral group (finite or infinite), with $S = \{s, t\}$. We claim that all $P_{x,w} = 1$ ($x \leq w$). The idea is to use induction on $\ell(w)$ together with (23). This is feasible because of the very simple behavior of the Bruhat ordering here: $x < w$ if and only if $\ell(x) < \ell(w)$. For each $w \neq 1, s, t$ there are precisely two elements of length $\ell(w) - 1$, one each with a reduced expression starting with s (resp. t). In the final sum in (23), at most one of these two occurs, and by induction no other term occurs: $z \prec v$ only if $\ell(v) - \ell(z) = 1$. Then the formula typically reads either $P_{x,w} = 1$ or $P_{x,w} = 1 + q - q$. (The reader should check the details.)

(b) The first rank 3 group which yields polynomials different from 1 is the group S_4. Let $S = \{s_1, s_2, s_3\}$, where $s_1 = (12), s_2 = (23), s_3 = (34)$. There are just two interesting cases, with $x \prec w$ but $\ell(w) - \ell(x) - 1 > 0$: $s_2 \prec s_2 s_1 s_3 s_2$ and $s_1 s_3 \prec s_1 s_3 s_2 s_3 s_1$. In both cases $P_{x,w} = 1 + q$.

(c) Alvis [1] has reported the computation of all the polynomials for the non-crystallographic finite Coxeter group of type H_4, and has thereby verified in particular that their coefficients are always nonnegative. He observes that there are 75 539 433 pairs $x \leq w$ (but substantial reductions can be made before the main computation is done).

(d) Dyer [2] has computed the polynomials for the 'universal' Coxeter group W having all $m(s, t) = \infty$ (for $s \neq t$). Again the coefficients are seen to be nonnegative.

(e) Lascoux–Schützenberger [1] have given an explicit combinatorial description of some of the Kazhdan–Lusztig polynomials for symmetric groups (in the 'grassmannian' case). (See Boe [1] for generalizations.)

(f) Goresky [1] has given the results of computer calculations for a number of finite reflection groups of small rank: types A_3, A_4, A_5, C_3, C_4, D_4, H_3. His tables give complete information about the Betti numbers and intersection homology dimensions for Schubert varieties in the flag variety of a semisimple group having W as Weyl group (although the results for the non-crystallographic group of type H_3 are just formal). His computer program uses the version of the Kazhdan–Lusztig algorithm developed by Gelfand–MacPherson [1]. He has also done calculations for affine Weyl groups.

(g) An appendix to Boe–Collingwood [1] contains a Fortran program developed by Boe for the computation of Kazhdan–Lusztig polynomials together with other data of representation-theoretic interest. More recently, F. du Cloux has created an interactive program *Coxeter*, allowing the user to compute Kazhdan-Lusztig polynomials and related data for Weyl groups of rank ≤ 6.

(h) As mentioned earlier, it is conjectured in Kazhdan–Lusztig [1] that all coefficients of $P_{x,w}$ are nonnegative. This has been verified in some of the most important special cases, where the coefficients can be computed explicitly (as in Alvis [1] and Dyer [2]) or can be interpreted as dimensions of cohomology groups (as in Kazhdan–Lusztig [2] for Weyl groups and affine Weyl groups, Haddad [1] for Coxeter groups which arise as 'Weyl groups' of Kac–Moody Lie algebras). Deodhar [11] has proposed a closed formula for the polynomials (described combinatorially), similar in spirit to his formula for R-polynomials (7.5). This will be correct if and only if the nonnegativity conjecture is true.

7.13   Inverse Kazhdan–Lusztig polynomials

Like the R-polynomials, the Kazhdan–Lusztig polynomials form an upper triangular unipotent matrix (infinite if W is infinite) relative to a total ordering of W compatible with the Bruhat ordering. Any such matrix can in principle be inverted over the ring A. For the R-polynomials, the inverse matrix was described in part (c) of Proposition 7.8; its entries are (up to sign) R-polynomials. For the $P_{x,w}$, nothing quite so simple can be written down in general, unless W is finite:

Proposition   *If W is finite, with longest element $w_\circ$, then for all $x \leq w$ we have:*

$$\sum_{x \leq z \leq w} \varepsilon_w \varepsilon_z P_{x,z} P_{w_\circ w, w_\circ z} = \delta_{x,w}. \tag{24}$$

Proof. This is clear when $x = w$. Proceed by induction on $\ell(w) - \ell(x)$, assuming $x < w$. If $D_{x,w}$ denotes the sum on the left side of (24), we just have to show that $D_{x,w} = 0$. The strategy is to introduce R-polynomials into the picture, via (20) in 7.10, then use the inversion formula for them, together with induction and the fact that $R_{y,z} = R_{w_\circ z, w_\circ y}$ for all $y \leq z$ (Proposition 7.6). First we write down two special cases of equation (20) in 7.10:

$$\bar{P}_{x,z} = q_z^{-1} q_x \sum_{x \leq u \leq z} R_{x,u} P_{u,z}$$

$$\bar{P}_{w_\circ w, w_\circ z} = q_{w_\circ z}^{-1} q_{w_\circ w} \sum_{w_\circ w \leq w_\circ v \leq w_\circ z} R_{w_\circ w, w_\circ v} P_{w_\circ v, w_\circ z}.$$

Using the fact that $\ell(w_\circ w) = \ell(w_\circ) - \ell(w)$, together with Proposition 7.6, the right side of the second equation can be simplified to:

$$q_z q_w^{-1} \sum_{z \leq v \leq w} R_{v,w} P_{w_\circ v, w_\circ z}.$$

Now substitute both equations into $\bar{D}_{x,w}$:

$$\bar{D}_{x,w} = \sum_{x \le z \le w} \varepsilon_w \varepsilon_z \sum_{x \le u \le z} \sum_{z \le v \le w} q_w^{-1} q_x R_{x,u} R_{v,w} P_{u,z} P_{w_o v, w_o z}.$$

Note that the product of P-polynomials here is the same one occurring in $D_{u,v}$, which we can substitute to obtain:

$$\bar{D}_{x,w} = \sum_{x \le u \le v \le w} \varepsilon_w \varepsilon_v q_w^{-1} q_x R_{x,u} R_{v,w} D_{u,v}.$$

By induction, $D_{u,v} = 0$ for all $u < v$ satisfying $\ell(v) - \ell(u) < \ell(w) - \ell(x)$. Inspection of the sum shows that only two kinds of terms survive (for the extremes $u = x, v = w$ and $u = v$):

$$\bar{D}_{x,w} = q_w^{-1} q_x D_{x,w} + q_w^{-1} q_x \sum_{x \le u \le w} \varepsilon_w \varepsilon_u R_{x,u} R_{u,w}.$$

Since $x < w$, part (c) of Proposition 7.8 shows that the last sum vanishes. (As noted there, the result is the same if ε_w is written in place of ε_x.) Thus:

$$\bar{D}_{x,w} = q_w^{-1} q_x D_{x,w}, \text{ or}$$

$$q_w^{\frac{1}{2}} q_x^{-\frac{1}{2}} \bar{D}_{x,w} = q_w^{-\frac{1}{2}} q_x^{\frac{1}{2}} D_{x,w}.$$

The degree bounds on the P-polynomials show that this ι-invariant expression is a polynomial in $q^{\frac{1}{2}}$ without constant term, forcing it to be 0 as desired. $\square$

Exercise. Recall the pairing μ introduced in 7.11: if $x < w$, $\mu(x, w)$ is the coefficient of $q^{(\ell(w) - \ell(x) - 1)/2}$ in $P_{x,w}$. If W is finite, prove that $\mu(x, w) = \mu(w_o w, w_o x)$ for all $x < w$. [We may assume $\varepsilon_x = -\varepsilon_w$ and $\varepsilon_{w_o x} = -\varepsilon_{w_o w}$. Rewrite the inversion formula:

$$P_{w_o w, w_o x} - P_{x,w} = \sum_{x < z < w} \varepsilon_x \varepsilon_z P_{x,z} P_{w_o w, w_o z}.$$

The key power of q does not occur on the right, so must occur with the same coefficient in each term on the left.]

When we set $q = 0$, the formula has an interesting consequence for the Bruhat ordering, to be explained in 8.5 below. Recall the exercise at the end of 7.11, which asserts that $P_{x,w}(0) = 1$ whenever $x \le w$. (This is easy to obtain by induction from formula (23) in that section.) Noting that the inversion formula is unchanged if we replace ε_w by ε_x, we get by substituting $q = 0$:

Corollary *For all $x \leq w \in W$,*

$$\sum_{x \leq z \leq w} \varepsilon_x \varepsilon_z = \delta_{x,w} = \sum_{x \leq z \leq w} \varepsilon_w \varepsilon_z.$$

In particular, if $x < w$, $\sum_{x \leq z \leq w} \varepsilon_z = 0$, i.e., the closed interval $[x, w]$ contains equally many elements of odd and of even length. □

Remark. Lusztig [2] has derived an explicit algorithm for inverting the polynomials $P_{x,w}$ in the case of affine Weyl groups, cf. Andersen [1], Kaneda [1], Kato [2]. The inverse polynomials are conjectured to have a nice interpretation in the modular representation theory of semisimple algebraic groups.

7.14   Multiplication formulas

In the program of Kazhdan–Lusztig [1], it is essential to see how the elements T_s act on the new basis $\{C_w | w \in W\}$ of $\mathcal{H}$. The precise answer involves the coefficients $\mu(x, w)$ and resulting relation $x \prec w$ introduced in the first paragraph of 7.11, together with equation (22) in that section.

Proposition *Let $s \in S, w \in W$.*
  (a) *If $sw < w$, then $T_s C_w = -C_w$.*
  (b) *If $w < sw$, then $T_s C_w = q C_w + q^{\frac{1}{2}} C_{sw} + q^{\frac{1}{2}} \sum \mu(z, w) C_z$, where the sum is taken over all $z \prec w$ for which $sz < z$.*

Proof. To prove (b), rewrite equation (22), with w in place of v (and sw in place of w):

$$C_{sw} = C_s C_w - \sum \mu(z, w) C_z,$$

where the sum is taken over $z \prec w$ for which $sz < z$. Substitute

$$C_s = q^{-\frac{1}{2}} T_s - q^{\frac{1}{2}} T_1$$

and rewrite with $T_s C_w$ on the left side to obtain (b).

  Next consider (a), assuming $sw < w$, so $\ell(w) \geq 1$. If $\ell(w) = 1$, we must have $w = s$ and we can check directly:

$$
\begin{aligned}
T_s C_s &= q^{-\frac{1}{2}} T_s^2 - q^{\frac{1}{2}} T_s \\
&= q^{-\frac{1}{2}} ((q-1) T_s + q T_1) - q^{\frac{1}{2}} T_s \\
&= -q^{-\frac{1}{2}} T_s + q^{\frac{1}{2}} T_1 \\
&= -C_s.
\end{aligned}
$$

Proceed by induction on $\ell(w)$. Note that part (b) can be applied to the situation $s(sw) > sw$:

$$T_s C_{sw} = q C_{sw} + q^{\frac{1}{2}} C_w + q^{\frac{1}{2}} \sum \mu(z, sw) C_z, \text{ or}$$

$$C_w = q^{-\frac{1}{2}}T_s C_{sw} - q^{\frac{1}{2}}C_{sw} - \sum \mu(z, sw)C_z.$$

Since $sz < z < w$ for each z involved in the sum, induction shows that $T_s C_z = -C_z$. Therefore

$$
\begin{aligned}
T_s C_w &= q^{-\frac{1}{2}}[(q-1)T_s + qT_1]C_{sw} - q^{\frac{1}{2}}T_s C_{sw} + \sum \mu(z, sw)C_z \\
&= q^{\frac{1}{2}}C_{sw} - q^{-\frac{1}{2}}T_s C_{sw} + \sum \mu(z, sw)C_z \\
&= -C_w. \quad \square
\end{aligned}
$$

Corollary *Let $x < w$. If $sw < w$ but $sx > x$ for some $s \in S$, then $P_{x,w} = P_{sx,w}$.*

Proof. By part (a) of the proposition, $T_s C_w = -C_w$. Compare the coefficient of T_{sx} on each side, using the formula for C_w in Theorem 7.9. Thanks to Lemma 7.4, $sx \leq w$, so T_{sx} does occur on the right side. On the left side, T_{sx} arises from the two terms in C_w involving T_x and T_{sx}. A quick calculation yields the corollary. $\square$

There are of course 'right-handed' versions of the proposition and corollary, which the reader can easily work out.

Exercise. Let W be finite, with longest element w_o. Prove that $P_{x,w_o} = 1$ for all $x \in W$. [Use the corollary repeatedly.]

7.15 Cells and representations of Hecke algebras

A central goal of Kazhdan–Lusztig [1] is to understand the representations of Hecke algebras. The formulas in Proposition 7.14 show how $\mathcal{H}$ acts on itself in the (left) regular representation, relative to the C-basis. But $\mathcal{H}$ is still a very large module, so one looks for smaller submodules (or subquotients). The advantage of the C-basis is that it leads to a systematic construction of representations associated with sets ('cells') which partition W. The description of cells is quite subtle, and combinatorially difficult to make explicit, but the applications in Lie theory have amply motivated this approach (see the references below).

Recall that we write $x \prec w$ if $x < w$ and the degree of $P_{x,w}$ is as large as possible: $(\ell(w) - \ell(x) - 1)/2$. Write $x{-}w$ if either $x \prec w$ or $w \prec x$. Next define subsets of S for each $w \in W$ by

$$L(w) := \{s \in S | sw < w\}, \quad R(w) := \{s \in S | ws < w\}.$$

For example, $L(1) = \emptyset = R(1)$ and (if W is finite, with longest element w_o) $L(w_o) = S = R(w_o)$. Note that we get crude partitions of W by calling elements w, w' equivalent if $L(w) = L(w')$ (resp. $R(w) = R(w')$).

Now define $x \leq_L w$ to mean that there is a chain $x = x_0, x_1, \ldots, x_r = w$ such that $x_i\text{—}x_{i+1}$ and $L(x_i)$ is not included in $L(x_{i+1})$ for $0 \leq i < r$. (There is a similar definition using R in place of L.) This transitive 'preorder' yields an equivalence relation on W: $x \sim_L w$ if and only if both $x \leq_L w$ and $w \leq_L x$ hold (or $x = w$). The resulting equivalence classes are called the **left cells** of W. There is an analogous definition of **right cells**.

We can also define $x \leq_{LR} w$ to mean that there exists a chain $x = x_0, x_1, \ldots, x_r = w$ such that for each $i < r$, either $x_i \leq_L x_{i+1}$ or $x_i \leq_R x_{i+1}$. This likewise yields an equivalence relation $x \sim_{LR} w$, whose equivalence classes are called the **two-sided cells** of W. It follows at once from the definitions that each two-sided cell is a union of left cells (resp. right cells).

Exercise. The identity element of W lies in a two-sided cell by itself. If W is finite, the same is true of w_0.

Example. Let $W = \mathcal{D}_m$, $m \leq \infty$. Say $S = \{s, t\}$. We saw in 7.12 that $P_{x,w} = 1$ whenever $x \leq w$ (i.e., whenever $\ell(x) \leq \ell(w)$). This implies at once that $x \prec w$ if and only if $\ell(w) - \ell(x) = 1$. So $x\text{—}w$ if and only if the length difference is 1. Now each element (other than w_0 if $m < \infty$) has a unique reduced expression. Clearly $L(w) = \{s\}$ if this expression begins with s; otherwise $L(w) = \{t\}$. Excluding 1 (and w_0 if W is finite), the condition $L(x) \not\subset L(w)$ therefore amounts to the requirement that x and w begin differently. Now we can construct chains such as

$$s\text{—}ts\text{—}sts\text{—}tsts \cdots ,$$

showing that $x \leq_L w$ whenever x, w have reduced expressions ending in s. The reverse chain also meets the condition for $w \leq_L x$, so $x \sim_L w$. On the other hand, it is clear that we cannot construct such chains joining elements which end differently. So there are exactly three left cells if $m = \infty$, four left cells if $m < \infty$. Similarly, we get right cells by considering elements which begin with s (resp. t). It is now clear that all elements of W except 1 and w_0 lie in a single two-sided cell.

Shi [1], 1.7, discusses in detail a number of other examples, including the much more complicated (and more interesting) case of symmetric groups. Here the cells turn out to involve the *Robinson–Schensted correspondence*, which sets up a one-to-one correspondence between group elements and pairs of 'standard Young tableaux' of the same shape.

The following proposition shows that in general the decomposition of W into left cells is a refinement of the decomposition mentioned earlier into sets of elements having a common R-set.

Proposition *If $x \leq_L y$, then $R(x) \supset R(y)$. Thus, if $x \sim_L y$, we have $R(x) = R(y)$.*

Proof. It is enough to consider the case $x \text{—} y$, with $L(x) \not\subset L(y)$.

(a) Suppose $y \prec x$. Say $s \in L(x) \setminus L(y)$, so $sx < x$ but $sy > y$. By Corollary 7.14 (applied to the pair y, x in place of x, w), $P_{y,x} = P_{sy,x}$. We claim this forces $x = sy$. Otherwise, we would have

$$\deg P_{y,x} = \deg P_{sy,x} \leq \frac{1}{2}(\ell(x) - \ell(sy) - 1) < \frac{1}{2}(\ell(x) - \ell(y) - 1),$$

contrary to the assumption $y \prec x$. Now $x = sy$ and $y < x$ together imply $R(x) \supset R(y)$.

(b) Suppose $x \prec y$. We reach a contradiction by supposing that there exists some $s \in R(y) \setminus R(x)$. A 'right-handed' version of the argument in (a) shows that $y = xs$, which (in view of $x < y$) forces $L(y) \supset L(x)$, contrary to assumption. $\square$

With the language of cells in mind, we can reconsider what the formulas in Proposition 7.14 tell us about the left regular representation of $\mathcal{H}$. In case (b) of the proposition, we have $w < sw$, so that $w \prec sw$ with $L(sw) \not\subset L(w)$, implying $sw \leq_L w$. On the other hand, any element $z \prec w$ in the sum satisfies $sz < z$ for the given s, so $L(z) \not\subset L(w)$ (because $sw > w$). Thus $z \leq_L w$. In either case of the proposition, it follows that left multiplication by T_s takes C_w into the A-span of itself and various C_x for which $x \leq_L w$.

Now fix a left cell $Z \subset W$, and define $\mathcal{I}_Z$ to be the A-span of all C_w ($w \in Z$) together with all C_x for which $x \leq_L w$ ($w \in Z$). The preceding discussion shows that $\mathcal{I}_Z$ is a left ideal in $\mathcal{H}$. Let $\mathcal{I}'_Z$ be the span of those C_x for which $x \leq_L w$ for some $w \in Z$ but $x \notin Z$. Since $\leq_L$ is transitive, the definition of left cell implies that $\mathcal{I}'_Z$ is also a left ideal in $\mathcal{H}$, so the quotient $\mathcal{M}_Z := \mathcal{I}_Z/\mathcal{I}'_Z$ affords a representation of $\mathcal{H}$. It is not too hard to see that it has a free A-basis in natural one-to-one correspondence with the elements C_w, $w \in Z$.

Similarly, one can use right cells and two-sided cells to define right $\mathcal{H}$-modules and $\mathcal{H}$-bimodules. In Kazhdan–Lusztig [1], the notion of 'W-graph' is introduced to make it easier to visualize what is going on, and numerous examples are given. The study of all these cell representations (especially in the case of Weyl groups) has been an important and challenging problem.

By now there is a lot of literature on cells and their connections with Lie theory. We conclude with a quick survey. For Weyl groups, see Lusztig [5], Chapter 5, as well as his papers [4][8]. Much of this is motivated by the representation theory of finite groups of Lie type (see Carter [4]). Representations and W-graphs associated with cells are also discussed by Curtis [3], Garsia–McLarnan [1], Gyoja [1][2], Heck [1], Kerov [1]. The group of type H_4 is treated by Alvis [1], Alvis–Lusztig [1].

The cells of affine Weyl groups have been described in special cases by Bédard [1][2], Du [1][2], Lawton [1], Lusztig [6], Lusztig–Xi [1], Shi [1]–[3]. In numerous papers, Lusztig has gone deeply into the general features of cells for affine Weyl groups, with applications to the representations of p-adic groups. He proves for example that there are only finitely many one-sided (hence two-sided) cells, and sets up a one-to-one correspondence between two-sided cells and unipotent classes in an associated simple algebraic group.

Some hyperbolic Coxeter groups of rank 3 have also been studied by Bédard [1][3]: here there may be infinitely many left cells.

Chapter 8

Complements

In this final chapter, we survey (without proofs) some related topics which may stimulate the reader to do further reading in the extensive literature of Coxeter groups. These deal with such matters as the internal structure of the groups, their representations, and the Bruhat ordering. The selection of topics and the order of presentation are somewhat random, with no claims of balance or completeness intended. Unless otherwise stated, (W, S) denotes an arbitrary Coxeter system.

8.1 The Word Problem

As noted at the end of 5.13, the concrete action of W on a fundamental domain in the dual of the vector space V affording its geometric representation could in principle be used to test which words in the generating set S are equal to 1. But this is extremely cumbersome in practice. Even if programmed for a computer, serious round-off problems can be anticipated, since it is essential to decide whether certain calculated coefficients are strictly positive.

A more attractive method was devised by Tits [5]. It allows one to transform an arbitrary product of generators from S into a reduced expression by making only the most obvious types of modifications coming from the defining relations. Here is a brief description, in our own notation. (For a nice reformulation of Tits' arguments, see pages 49–52 of Brown [1].)

Let F be a free group on a set Σ in bijection with S (with σ corresponding to s), and let $\pi : F \to W$ be the resulting epimorphism. Since each $s^2 = 1$, the monoid F^+ generated by Σ already maps onto W. If $\omega \in F^+$ is a product of various elements σ, we can define $\ell(\omega)$ to be the number of factors involved. If $m = m(s, t)$ for $s, t \in S$, the product of m factors $\sigma\tau\sigma \cdots$ maps to the same element of W as the product of m

factors (in reverse order) $\tau\sigma\tau\cdots$. Replacement of one of them by the other inside a given $\omega \in F^+$ is called an 'elementary simplification' of the first kind; it leaves the length undisturbed. A second kind of elementary simplification reduces length, by omitting a consecutive pair $\sigma\sigma$. Write $\Sigma(\omega)$ for the set of all elements of F^+ obtainable from ω by a sequence of elementary simplifications. Since no new elements of Σ are introduced and length does not increase at each step, it is clear that $\Sigma(\omega)$ is finite. It is also effectively computable (though it may take some trouble to devise an efficient algorithm). Clearly the image of $\Sigma(\omega)$ under π is a single element of W.

Theorem *Let* $\omega, \omega' \in F^+$. *Then* $\pi(\omega) = \pi(\omega')$ *if and only if* $\Sigma(\omega)$ *meets* $\Sigma(\omega')$. *In particular,* $\pi(\omega) = 1$ *if and only if the empty word lies in* $\Sigma(\omega)$.

One direction is obvious. To go the other way, Tits assumes the contrary and analyzes a minimal counterexample (in terms of lexicographic ordering of pairs (ω, ω')): both elements must have the same length and $\Sigma(\omega)$ consists of elements of equal length, etc., leading eventually to a contradiction.

Much less seems to be known about the *Conjugacy Problem* for Coxeter groups: given $w, w' \in W$, decide whether or not they are conjugate. Appel–Schupp [1] have shown how to solve the problem for 'extra-large' Coxeter groups (those for which all $m(s, s') \geq 4$ when $s \neq s'$).

8.2   Reflection subgroups

In 5.7 we defined 'reflections' in W to be the conjugates of elements of S; denote by T the set of all reflections. Define a **reflection subgroup** of W to be any subgroup W' generated by a subset of T. Independently, Deodhar [10] and Dyer [3] have proved that reflection subgroups of W are also Coxeter groups. (When W 's finite, this is clear already from Chapter 1.) To formulate this precisely, one first has to pick out a suitable set of distinguished generators. Dyer proceeds as follows.

For any $w \in W$, set $N(w) := \{t \in T | \ell(tw) < \ell(w)\}$. This set is especially interesting when w itself lies in T. If a reflection $t' \in W'$ is to behave like a 'simple' reflection, it should not be possible to reduce its length by multiplying by another reflection t in W'. Accordingly, define S' to be the set of all $t' \in T$ for which $N(t') \cap W'$ consists of t' alone. In particular, $S' \subset W'$.

Theorem *Let* T *be the set of reflections in* W, *and let* W' *be a subgroup generated by some subset of* T. *Then, with* S' *defined as above,* (W', S') *is a Coxeter system.*

Dyer's idea is to view the function N (from W to subsets of T) as a 'cocycle', which behaves well on restriction to W'. His arguments in fact apply to a larger class of 'reflection systems', with generators not necessarily of order 2, which are characterized by the existence of such a cocycle. In terms of the geometric representation of W, he can also characterize the possible sets of canonical generators of reflection subgroups in terms of the angles between corresponding roots.

Deodhar's proof that W' is a Coxeter group relies instead on his earlier characterization of Coxeter groups in terms of properties of their root systems (Deodhar [4][7]). His idea is to locate a 'simple system' in the set of roots naturally associated to W', then define S' to be the corresponding set of reflections. Because his requirements on a root system are all met by these data, (W', S') must be a Coxeter system. (He also remarks that, according to Tits, another proof of the theorem can be based on Proposition 3 in Tits [9].)

8.3 Involutions

There does not appear to be any uniform way to describe or parametrize the conjugacy classes of an arbitrary Coxeter group. Even for Weyl groups, where the classes have been determined explicitly, a unified approach turns out to be quite difficult (see Carter [3]). But in the case of *involutions* (elements of order 2), a satisfying general method is given by Richardson [1].

We want to locate a small set of involutions to which all others must be conjugate. As in other investigations of the group-theoretic properties of W, the geometric representation (5.3) plays a major role here. Recall from 5.5 that, for a subset $I \subset S$, the geometric representation of the parabolic subgroup W_I can be realized as the action of W_I on the subspace V_I of V spanned by the roots α_s, $s \in I$. It may or may not be true that the operator $-1 \in \mathrm{GL}(V_I)$ lies in W_I: this would require that W_I be finite (thanks to part (b) of Proposition 5.6), and then that -1 be equal to the unique longest element (as determined for each type by Corollary 3.19). This always happens, of course, when $|I| = 1$. If $-1 \in W_I$, call it w_I and say that I satisfies the '(-1)-condition'. Denote by $\mathcal{J}$ the collection of all subsets of S satisfying the (-1)-condition. Finally, define $I, J \in \mathcal{J}$ to be 'W-equivalent' if some $w \in W$ maps $\{\alpha_s | s \in I\}$ onto $\{\alpha_s | s \in J\}$.

Theorem (a) *Each involution in W is conjugate to some w_I, $I \in \mathcal{J}$.*

(b) *For $I, J \in \mathcal{J}$, w_I is conjugate to w_J if and only if I and J are W-equivalent.*

While this gives a reasonably explicit description of the conjugacy classes of involutions, it takes some work to make it effective. Using techniques of Deodhar [4] and Howlett [1], Richardson formulates an algorithm for testing W-equivalence of subsets I, J (in terms of 'elementary' equivalences) and illustrates it for the affine Weyl group $\widetilde{E_7}$.

Besides parametrizing the conjugacy classes of involutions in an efficient way, one can study involutions by writing each as a product of commuting reflections. Deodhar [4] actually obtains a canonical decomposition of this sort: for each involution w there exists a unique 'totally orthogonal' set of roots such that w is the product of the corresponding (commuting) reflections. (See also Springer [5].)

8.4   Coxeter elements and their eigenvalues

Recall from 3.16–3.20 the discussion of Coxeter elements for a finite Coxeter group. Such elements form a single conjugacy class, and their eigenvalues turn out to determine in a very simple way the degrees of basic polynomial invariants for the group. Here we consider what can be said about an arbitrary (irreducible) Coxeter group.

Fix an enumeration $S = \{s_1, \ldots, s_n\}$, and define $w := s_1 \cdots s_n$ to be a **Coxeter element** of W. The proof of Proposition 3.16 shows that all such elements form a single conjugacy class if the Coxeter graph Γ is a tree. But this may fail when Γ contains circuits. Even so, the study of Coxeter elements yields some interesting dividends.

When W is finite, one could study the eigenvalues of a Coxeter element by looking directly at the corresponding matrix and its characteristic polynomial, as was done in Coxeter [4] (see Bourbaki [1], pp. 140–141). We actually used less direct methods, but the matrix approach has its advantages when W is no longer finite.

Howlett [2] gives the following description. Start with the geometric representation $\sigma : W \to \mathrm{GL}(V)$ as in (5.3) and let A denote the matrix (for a chosen ordering of the roots α_s) of the associated bilinear form B, whose values on the basis of V are given by

$$B(\alpha_s, \alpha_{s'}) = -\cos\frac{\pi}{m(s, s')}.$$

Write $2A = U + U^{\mathrm{t}}$, where U is an upper triangular unipotent matrix and U^{t} is its transpose. Then the matrix representing the Coxeter element w defined by the chosen ordering is shown by Howlett to be simply $-U^{-1}U^{\mathrm{t}}$. Moreover, U is an 'M-matrix' (having nonpositive off-diagonal entries, but positive principal minors), so the general theory of such matrices can be used to study the eigenvalues of $\sigma(w)$.

For example, when W is the affine Weyl group of type $\widetilde{A_1}$ (an infinite dihedral group), with $S = \{s_0, s_1\}$, $w = s_0 s_1$ is conjugate to $s_1 s_0$ (by

s_0), and the matrix representing w is

$$\begin{pmatrix} 3 & -2 \\ 2 & -1 \end{pmatrix} = - \begin{pmatrix} 1 & 2 \\ 0 & 1 \end{pmatrix} \begin{pmatrix} 1 & 0 \\ -2 & 1 \end{pmatrix}$$

Howlett is able to prove:

Theorem *W is infinite if and only if the Coxeter element w is of infinite order, if and only if $\sigma(w)$ has a real eigenvalue ≥ 1. Moreover, W is of affine type (with B positive semidefinite) if and only if 1 is an eigenvalue of $\sigma(w)$ and all other eigenvalues have absolute value 1.*

Earlier A'Campo [1] had shown, assuming Γ has no circuits, that W is infinite precisely when w has infinite order. His analysis of the spectrum of a Coxeter element (motivated by an application to singularity theory) began by showing, if Γ has no circuits, that the eigenvalues of $\sigma(w)$ are of absolute value 1 or else lie in $\mathbf{R}^+$. But this may fail when Γ contains a circuit, according to an example of Berman–Lee–Moody [1].

Coxeter elements for affine Coxeter groups, or more generally for the Weyl groups of Kac–Moody Lie algebras, have been studied recently by a number of people, including Berman–Lee–Moody [1], Coleman [2], Steinberg [6].

8.5 Möbius function of the Bruhat ordering

Next we survey some topics involving the Bruhat ordering of W (5.9). The basic references for this section are Verma [1][2], Deodhar [1], and Kazhdan–Lusztig [1].

We begin by recalling some general notions about a partially ordered set $(X, \leq)$ satisfying the condition: for each $y \in X$, the set $\{x \in X | x \leq y\}$ is finite. Define $I := \{(x, y) \in X \times X | x \leq y\}$. Then there is a unique function $\mu : I \to \mathbf{Z}$ (called the **Möbius function** of X) such that, for any $(x, y) \in I$,

$$\sum_{x \leq z \leq y} \mu(x, z) = \delta_{x,y} \text{ (Kronecker delta)}.$$

The summation could equally well be taken over all $\mu(z, y)$.

With the aid of μ one gets a Möbius inversion formula of the following type. Given any function f from X into an abelian group (written additively), set $g(x) := \sum_{y \leq x} f(y)$. Then

$$f(y) = \sum_{x \leq y} \mu(x, y) g(x).$$

This generalizes the familiar formula for the set of positive integers, partially ordered by the divisibility relation.

The Bruhat ordering of W satisfies the finiteness condition above, insuring the existence of a Möbius function μ. (*Warning:* This function μ should not be confused with the unrelated function defined in 7.11!) From the work of Kazhdan–Lusztig [1] we derived in Corollary 7.13 the following formula for $x \leq w$ when W is finite:

$$\sum_{x \leq z \leq w} \varepsilon_x \varepsilon_z = \delta_{x,w},$$

where $\varepsilon_x = (-1)^{\ell(x)}$. The uniqueness of the Möbius function then implies a simple formula for μ, which turns out to be true for all Coxeter groups:

Theorem   *If $x < w$ in W, then $\mu(x,w) = (-1)^{\ell(x)+\ell(w)}$.*

As noted in 7.13, if $x < w$, the formula implies that the number of elements of even length in the interval $[x, w]$ equals the number of elements of odd length. In particular, if $\ell(w) - \ell(x) = 2$, then there are precisely two elements z satisfying $x < z < w$.

The history of these results is a bit complicated. Apparently the result just mentioned on intervals of length 2 was first proved (for Weyl groups) by Bernstein–Gelfand–Gelfand [2] as part of their construction of the 'BGG resolution'; their argument can be generalized to cover all Coxeter groups. They noted that the result would also follow from Verma's then unpublished work giving the description of the Möbius function (as in the theorem), which later appeared as Verma [1]. The latter paper has a gap (in the fifth line of Case 2 in the lemma), which Verma pointed out in an 'Erratum' and then filled in an unpublished paper (Verma [2]). Subsequently Deodhar [1] proved a more comprehensive result on Möbius functions (formulated below). Then the paper of Kazhdan–Lusztig [1] provided an indirect determination of the Möbius function for finite W, as in our Corollary 7.13.

Deodhar [1] obtains a more general result, by considering the set W^I of minimal coset representatives for the parabolic subgroup W_I in W (5.12), with the partial ordering induced by the Bruhat ordering. His result on the Möbius function μ^I of this set may be formulated as follows: if $x < w$ in W^I, then $\mu^I(x,w) = (-1)^{\ell(w)+\ell(x)}$ if the full interval $[x, w]$ in W lies in W^I, but otherwise $\mu^I(x,w) = 0$. When I is empty, we recover the above theorem.

8.6   Intervals and Bruhat graphs

As noted in 8.5, intervals of length 2 in the Bruhat ordering contain exactly two intermediate elements. Longer Bruhat intervals are potentially much more complicated, when viewed as partially ordered sets (posets) in their own right. But in answer to a question of Björner [2], it is shown by Dyer [4] that

Theorem *For finite Coxeter groups, only finitely many isomorphism types of posets of a fixed length can occur as Bruhat intervals.*

To prove this and related results, Dyer introduces a 'Bruhat graph' to capture more refined information about the Bruhat ordering. Its vertices are the elements of W, with an edge directed from tw to w if t is a reflection (as in 8.2) and $\ell(tw) < \ell(w)$. Normally one would picture the Bruhat ordering by joining such vertices only when the length difference is exactly 1. But this more detailed graph behaves better with respect to inclusions of reflection subgroups.

For a closed Bruhat interval Z, Dyer shows that the isomorphism type of the poset Z completely determines the isomorphism type of the full subgraph of the Bruhat graph with vertex set Z. This provides some support for his conjecture (originating in a question of Kazhdan–Lusztig) that the polynomials $R_{x,w}$ of 7.4 and hence the Kazhdan–Lusztig polynomials $P_{x,w}$ (7.9) should depend only on the isomorphism type of the Bruhat interval $[x, w]$.

8.7 Shellability

If $x < w$, Proposition 5.11 implies that all maximal chains from x to w have the same length $\ell(w) - \ell(x)$. Maximal chains behave even more nicely, from a combinatorial viewpoint, as formulated by Björner–Wachs [1]:

Theorem *Any closed interval $[x, w]$ in the Bruhat ordering of W is lexicographically shellable.*

Their approach works more generally for intervals in W^I (or in 'descent classes'; see Björner [2]). Rather than explain the precise meaning of 'lexicographic shellability' for a poset, we shall formulate the essential idea for an interval $[x, w]$ in W. Fix a reduced decomposition $w = s_1 \cdots s_r$, and suppose $\ell(w) - \ell(x) = q$. Consider a maximal chain

$$x = w_0 < w_1 < \ldots < w_q = w.$$

From the Strong Exchange Condition one sees that the w_i may be obtained from w by systematic removal of one (uniquely determined) s_i at a time. So one can associate a 'label' to the maximal chain, consisting of the sequence of subscripts i (taken in the order of removal of the corresponding s_i). It can be shown that there is a unique maximal chain whose label is an increasing sequence, and moreover this label comes earlier in the lexicographic ordering of sequences than any other label of a maximal chain. This is (roughly) the content of the theorem.

As a consequence of the theorem, one recovers the formula for the Möbius function of the Bruhat ordering described above in 8.5. There is also an attractive topological formulation of shellability (Björner [2][3]), when one passes from a poset to its order complex: the simplicial complex whose simplices are the chains in the poset. As a corollary of shellability, one finds (for example) that the associated Stanley–Reisner ring is Cohen–Macaulay.

For another proof of the Björner–Wachs theorem, see Deodhar [5], §6.

8.8   Automorphisms of the Bruhat ordering

Viewing W just as a partially ordered set (relative to the Bruhat ordering), we can ask for a description of its automorphisms: bijections $\theta : W \to W$ for which $x < w$ if and only if $\theta(x) < \theta(w)$. Since 1 is the unique minimal element, $\theta(1) = 1$.

One obvious example is the inversion map $\theta(w) = w^{-1}$. Another type of automorphism arises whenever the Coxeter graph admits an automorphism (a bijection of vertices preserving labels on edges): such a graph automorphism respects the defining relations of W and hence induces a unique group automorphism, which in turn clearly preserves the Bruhat ordering as characterized in 5.10. Note that if the graph has more than one connected component of the same type, a permutation of these components is an example of a graph automorphism.

Thanks to Proposition 5.11, $\ell(w)$ may be characterized as the length of a maximal chain from 1 to w in the Bruhat ordering, so any automorphism θ preserves lengths. In particular, $\theta(S) = S$. In turn, it follows from 5.10 that, for any pair $s \neq t$ in S, $w \in W_{\{s,t\}}$ if and only if $\theta(w) \in W_{\{\theta(s),\theta(t)\}}$. Comparing orders, we see that the restriction of θ to S induces a graph automorphism.

When $|S| = 2$, $W = \mathcal{D}_m$ (where $m \leq \infty$) and the Bruhat ordering is uncomplicated: $x < w$ if and only if $\ell(x) < \ell(w)$, and there are precisely two elements of each nonzero length $< m$. Thus the automorphisms of the ordering correspond to all possible interchanges of elements of equal length, and the automorphism group is isomorphic to $(\mathbf{Z}/2\mathbf{Z})^{m-1}$. It turns out that this case is misleading, however. In higher ranks the possibilities are much more limited:

Theorem   *Let $|S| \geq 3$, and assume (W, S) is irreducible. Then every automorphism of the Bruhat ordering is a graph automorphism or else the composite of inversion with a graph automorphism.*

This was proved by van den Hombergh [1] (and recently rediscovered by Waterhouse [1]). In case W has more than one irreducible component, the statement of the theorem becomes slightly more complicated, since θ might act as inversion on only some of the components (and W may have dihedral factors as well).

8.9 Poincaré series of affine Weyl groups

Recall from 3.15 the factorization of the Poincaré polynomial $W(t) = \sum_w t^{\ell(w)}$ for a finite Coxeter group:

$$W(t) = \prod_{i=1}^{n} \frac{t^{d_i} - 1}{t - 1},$$

where $d_1, \ldots, d_n$ are the degrees of basic polynomial invariants of W. We followed Steinberg's version of Solomon's proof (using the Coxeter complex), but the earlier proof for Weyl groups by Chevalley [3] was based instead on the topology of a compact Lie group G having W as Weyl group (Chevalley [1]): G has the same cohomology as a product of spheres of dimensions $2d_i - 1$, and therefore its Poincaré polynomial factors as

$$\prod (1 + t^{2d_i - 1}).$$

From this one can derive (by a spectral sequence argument) the Poincaré series of the 'loop space' ΩG:

$$\prod (1 - t^{2(d_i - 1)})^{-1}.$$

Using some Morse theory, Bott [1] worked out a cell decomposition of ΩG indexed by the coroot lattice $L(\Phi^\vee)$ of W, thereby showing that the Poincaré series of $L(\Phi^\vee)$ (viewed as a subgroup of the affine Weyl group W_a) has the same form as that of ΩG, but with t replaced by its square root to take into account the difference between real and complex dimensions. (See 5.12 for the general notion of Poincaré series of a Coxeter subgroup or subset thereof.)

These results can be fitted together, using the fact that W is a parabolic subgroup W_I of W_a (where I corresponds to the Coxeter graph of W contained in the Coxeter graph of W_a). The distinguished coset representatives W^I are then in bijection with the coroot lattice, and we can obtain the Poincaré series $W_a(t)$ by multiplying the two series together as in 5.12. The end result is Bott's theorem:

Theorem

$$W_a(t) = W(t) \prod \frac{1}{1 - t^{m_i}},$$

where the $m_i = d_i - 1$ are the exponents of W (3.19).

See Bott [2] and Hiller [3], V.6, for further details of Bott's method, as well as later interpretations involving generalized Bruhat cells. (See also Iwahori–Matsumoto [1], 1.10.) Steinberg [5], §3, gives a more self-contained combinatorial proof of the theorem (in a more general 'twisted' version), and Macdonald [2], §3, works out another expression for the Poincaré series in terms of heights of roots (cf. 3.20).

8.10   Representations of finite Coxeter groups

Since the early work of Frobenius and Schur on representations of finite groups, symmetric and dihedral groups have served as natural examples. It is not difficult to work out explicitly the irreducible representations (over $\mathbf{C}$) of a dihedral group (see p. 339 of Curtis–Reiner [1]); their degrees are all 1 or 2. The natural two-dimensional 'reflection representation' of course figures in the list. Representations of symmetric groups are more complicated to construct, but these too have been rather completely worked out, in a combinatorial spirit: partitions of n parametrize both the classes and the irreducible representations of $\mathcal{S}_n$ (Curtis–Reiner [1], §28).

One remarkable feature of the representation theory of symmetric groups is the fact that all of their irreducible representations can be realized over $\mathbf{Q}$, i.e., the representing matrices can be chosen to have rational entries. It follows that the character values (traces) lie in $\mathbf{Z}$, since they are in any case algebraic integers (sums of roots of unity). This is not true for most dihedral groups. To see this, recall the discussion of crystallographic groups in 2.8. The proof of Proposition 2.8 shows that if all traces lie in $\mathbf{Z}$, then the possible values of $m(s, s')$ are limited to 1, 2, 3, 4, 6. Similarly, the characters of the group of type H_3 are readily found (from those of the alternating group of order 60), and are not all $\mathbf{Z}$-valued.

What can be said about representations of an arbitrary finite reflection group W (over $\mathbf{C}$ or its subfields)? The groups of types B_n and D_n are close relatives of symmetric groups and have been studied in the same explicit manner. Actual representations of the exceptional groups are much harder to come by, but the individual character tables have all been computed: see Frame [1] for the groups of type E_6, E_7, E_8, Kondo [1] for F_4, Grove [1] for H_4. It turns out that all characters of Weyl groups are $\mathbf{Z}$-valued, whereas this fails for non-crystallographic groups such as H_4. Even more is true, but requires nontrivial arguments for the exceptional types (Benard [1]):

Theorem *All irreducible representations of Weyl groups can be realized over* **Q**.

The distinction between having traces in a given field and having all matrix entries in that field is a subtle one, measured by the 'Schur index'. (See Kletzing [1], Chapter 5 and Appendix, for a discussion of exceptional Weyl groups, including character tables. For type H_4, see Benson–Grove [2].)

The theorem was originally proved in case-by-case fashion, with no unifying idea involved. A more sophisticated method of constructing Weyl group representations (by making W act on rational cohomology groups of certain algebraic varieties arising from an algebraic group having W as Weyl group) yields a more uniform proof (Springer [4], Corollary 1.15), though there is still some mild case-by-case analysis of the algebraic groups. The paper of Kazhdan–Lusztig [1] on representations of Hecke algebras and Coxeter groups was partly motivated by Springer's work, and achieved an even broader perspective. When W is a symmetric group, their work on cells (sketched in 7.15 above) yields a canonical basis over **Q** for each irreducible representation of W (see also Garsia–McLarnan [1]). But in general the actual representations they construct are not irreducible, a feature also of Springer's cohomology construction. See Carter [4], Chapters 11, 12 (and especially 12.4–12.6) for an account of the way these ideas interact with the study of characters of finite groups of Lie type.

8.11 Schur multipliers

To any group G is associated its **Schur multiplier**, which arose originally in connection with the lifting of projective representations to covering groups (Curtis–Reiner [1], §53). This is defined traditionally as the set of equivalence classes of factor sets $G \times G \to \mathbf{C}^*$, made into a group via pointwise multiplication of factor sets. In more modern notation, it becomes the cohomology group $H^2(G, \mathbf{C}^*)$. Some people instead define the Schur multiplier to be the homology group $H_2(G, \mathbf{Z})$, which can be computed from a presentation $G = F/N$ (F free) as $((F,F) \cap N)/(F,N)$. In general,
$$H^2(G, \mathbf{C}^*) \cong \mathrm{Hom}(H_2(G, \mathbf{Z}), \mathbf{C}^*).$$
We adopt the cohomology version here.

Following earlier work of Ihara–Yokonuma [1] and Yokonuma [1] on finite and affine Coxeter groups (see Karpilovsky [1], 7.2–7.3), Howlett [3] determines the Schur multiplier of an arbitrary Coxeter group. His description involves the computation of some invariants attached to the Coxeter graph Γ of W, as follows. Let n_1 be the rank (the number of vertices of Γ). Let n_2 be the number of edges of Γ having finite labels

$3 \leq m < \infty$. Let n_3 be the number of equivalence classes in the set of pairs of non-adjacent vertices of Γ, for the equivalence relation generated by:

$$\{s, s'\} \approx \{s, s''\} \quad \text{if } m(s', s'') \text{ is odd}.$$

Finally, let n_4 be the number of connected components of the graph obtained from Γ by deleting all edges whose labels are even or ∞.

Theorem *The Schur multiplier of W is an elementary abelian 2-group of rank $n_2 + n_3 + n_4 - n_1$.*

While the statement of the theorem is straightforward, the proof is quite intricate; Howlett works in a very explicit way with the presentation of W.

See Maxwell [3] for the Schur multiplier of the subgroup W^+ when W is finite.

8.12   Coxeter groups and Lie theory

A major impetus for the study of Coxeter groups has been their connections with semisimple Lie theory. Indeed, this is the *raison d'être* of Bourbaki [1]. In this brief concluding section, we have to be content with pointing the reader toward a small sample of the relevant literature.

Weyl groups of simple Lie groups and Lie algebras play a pervasive role in both structure theory and representation theory, beginning with the work of W. Killing and E. Cartan (and, somewhat later, Weyl). Consult Witt [1], Humphreys [1], and the other chapters of Bourbaki's treatise, or any of the vast number of books on Lie groups and their applications. For the topology of Lie groups, see Chevalley [1]. Recent work on representations (often infinite-dimensional) gives an even more prominent role to the Weyl group and associated Hecke algebra: see for example Bernstein–Gelfand–Gelfand [2], Deodhar [3][5][8][11], Gelfand–MacPherson [1], Jantzen [1][2], Kazhdan–Lusztig [1][2], Springer [6], Vogan [1].

Affine Weyl groups also play a major role in the study of compact Lie groups, by E. Cartan, A. Borel and J. de Siebenthal, and others; see Stiefel [1], Bott [2], Bröcker–tom Dieck [1].

The study of p-adic Lie groups is more recent. Here the affine Weyl group plays a new structural role, starting with Iwahori–Matsumoto [1] (see the exposition in Brown [1]) and culminating in the general theory of Bruhat–Tits. See also Lusztig [3], Macdonald [1], for some of the associated representation theory.

Semisimple algebraic groups over fields of prime characteristic share much of the structure of semisimple Lie groups, again depending heavily on the Weyl group; see Humphreys [2]. In recent years affine Weyl groups and Hecke algebras have also become prominent in the study of representations of semisimple groups in prime characteristic, though the reason for this is only imperfectly understood: see Jantzen [3], Lusztig [1][2], Verma [3]. All of this in turn has strong implications for the study of finite groups of Lie type, including their ordinary and modular representations: see Carter [1][2][4], Chevalley [3], Curtis [1]–[3], Curtis–Reiner [3], Iwahori [1], Solomon [3], Steinberg [4].

Kac–Moody Lie algebras (and associated groups) generalize the entire theory of semisimple Lie algebras (and Lie groups), and involve 'Weyl groups' which may be arbitrary crystallographic Coxeter groups: see for example Kac [1], Macdonald [3][5]. This part of Lie theory has grown rapidly in recent years, because of its many connections with theoretical physics, combinatorics, modular functions, etc.

Through much of this work runs a geometric thread: *Schubert varieties*. Classically, these occur as closures of Bruhat cells, which figure in the Bruhat decomposition of a semisimple group and are parametrized by the Weyl group. Inclusions of Bruhat cells in the closures of others are governed precisely by the Bruhat ordering of the Weyl group. This theme recurs in *p*-adic groups and Kac–Moody groups. The work of Kazhdan–Lusztig [1][2] makes a profound connection between the geometry of Schubert varieties and the representation theory of semisimple Lie algebras. Similar connections are found (or predicted) elsewhere in Lie theory.

'... si les choses se répètent, c'est avec de grandes variations'
Proust, *La Prisonnière*

References

N. A'Campo
1. Sur les valeurs propres de la transformation de Coxeter, *Invent. Math.* **33** (1976), 61–67.

E. Akyildiz, J.B. Carrell
1. A generalization of the Kostant–Macdonald identity, *Proc. Natl. Acad. Sci. U.S.A.* **86** (1989), 3934–3937.

D. Alvis
1. The left cells of the Coxeter group of type H_4, *J. Algebra* **107** (1987), 160–168.

D. Alvis, G. Lusztig
1. The representations and generic degrees of the Hecke algebra of type H_4, *J. Reine Angew. Math.* **336** (1982), 201–212.

H.H. Andersen
1. An inversion formula for the Kazhdan–Lusztig polynomials for affine Weyl groups, *Adv. in Math.* **60** (1986), 125–153.

K.I. Appel, P.E. Schupp
1. Artin groups and infinite Coxeter groups, *Invent. Math.* **72** (1983), 201–220.

H. Asano
1. A remark on the Coxeter–Killing transformations of finite reflection groups, *Yokohama Math. J.* **15** (1967), 45–49.

R. Bédard
1. Cells for two Coxeter groups, *Comm. Algebra* **14** (1986), 1253–1286.
2. The lowest two-sided cell for an affine Weyl group, *Comm. Algebra* **16** (1988), 1113–1132.
3. Left V-cells for hyperbolic Coxeter groups, *Comm. Algebra* **17** (1989), 2971–2997.

M. Benard
1. On the Schur indices of characters of the exceptional Weyl groups, *Ann. of Math.* **94** (1971), 89–107.

C.T. Benson, L.C. Grove

1. *Finite Reflection Groups*, Bogden & Quiqley, Tarrytown-on-Hudson, NY, 1971.
2. The Schur indices of the reflection group I_4, *J. Algebra* **27** (1973), 574–578.

S. Berman, Y.S. Lee, R.V. Moody

1. The spectrum of a Coxeter transformation, affine Coxeter transformations, and the defect map, *J. Algebra* **121** (1989), 339–357.

I.N. Bernstein, I.M. Gelfand, S.I. Gelfand

1. Schubert cells and cohomology of the spaces G/P, *Russian Math. Surveys* **28** (1973), 1–26.
2. Differential operators on the base affine space and a study of g-modules, *Lie groups and their representations*, Halsted, New York, 1975, pp. 21–64.

W.M. Beynon, G. Lusztig

1. Some numerical results on the characters of exceptional Weyl groups, *Math. Proc. Cambridge Philos. Soc.* **84** (1978), 417–426.

A. Björner

1. Some combinatorial and algebraic properties of Coxeter complexes and Tits buildings, *Adv. in Math.* **52** (1984), 173–212.
2. Orderings of Coxeter groups, *Combinatorics and Algebra*, Contemporary Math. vol. 34 , Amer. Math. Soc., 1984, pp. 175–195.
3. Posets, regular CW complexes and Bruhat order, *European J. Combin.* **5** (1984), 7–16.
4. Coxeter groups and combinatorics, *Proc. 19th Nordic Congress of Mathematicians* (Reykjavik, 1984), Icel. Math. Soc., Reykjavik, 1985, pp. 24–32.

A. Björner, M. Wachs

1. Bruhat order of Coxeter groups and shellability, *Adv. in Math.* **43** (1982), 87–100.
2. Generalized quotients in Coxeter groups, *Trans. Amer. Math. Soc.* **308** (1988), 1–37.

B.D. Boe

1. Kazhdan–Lusztig polynomials for Hermitian symmetric spaces, *Trans. Amer. Math. Soc.* **309** (1988), 279–294.

B.D. Boe, D.H. Collingwood

1. Multiplicity free categories of highest weight representations I, *Comm. Algebra* **18** (1990), 947–1032.

R. Bott

1. An application of the Morse theory to the topology of Lie-groups, *Bull. Soc. Math. France* **84** (1956), 251–281.

2. The geometry and representation theory of compact Lie groups, *Representation Theory of Lie Groups*, London Math. Soc. Lecture Note Ser. **34**, Cambridge University Press, 1979, pp. 65–90.

N. Bourbaki
1. *Groupes et algèbres de Lie*, Ch. 4–6, Hermann, Paris, 1968; Masson, Paris, 1981.

E. Brieskorn
1. Sur les groupes de tresses (d'après V.I. Arnol'd), *Sém. Bourbaki (1971/72)*, Exp. 401, Lect. Notes in Math. **317**, Springer, Berlin, 1973.
2. Die Fundamentalgruppe des Raumes der regulären Orbits einer endlichen komplexen Spiegelungsgruppe, *Invent. Math.* **12** (1971), 57–61.

E. Brieskorn, K. Saito
1. Artin-Gruppen und Coxeter-Gruppen, *Invent. Math.* **17** (1972), 245–271.

T. Bröcker, T. tom Dieck
1. *Representations of Compact Lie Groups*, Springer, New York, 1985.

N. Broderick, G. Maxwell
1. The crystallography of Coxeter groups II, *J. Algebra* **44** (1977), 290–318.

K.S. Brown
1. *Buildings*, Springer, New York, 1989.

J.B. Carrell
1. Some remarks on regular Weyl group orbits and the cohomology of Schubert varieties, *Contemp. Math.*, Amer. Math. Soc., to appear.

R.W. Carter
1. *Simple Groups of Lie Type*, J. Wiley & Sons, London, 1972.
2. Weyl groups and finite Chevalley groups, *Proc. Cambridge Philos. Soc.* **67** (1970), 269–276.
3. Conjugacy classes in the Weyl group, *Compositio Math.* **25** (1972), 1–59.
4. *Finite Groups of Lie Type: Conjugacy Classes and Complex Characters*, Wiley Interscience, London, 1985.

P. Cartier
1. Groupes finis engendrés par des symétries, Exposé 14, *Séminaire C. Chevalley 1956-1958*, Paris, 1958.

R. Charney, M. Davis
1. Reciprocity of growth functions of Coxeter groups, *Geom. Dedicata* **39** (1991), 373–378.

M. Chein
  1. Recherche des graphes des matrices de Coxeter hyperboliques
     d'ordre $\leq$ 10, *Rev. Francaise Informat. Recherche Opérationnelle*
     3 (1969), Sér. R-3, 3–16.

C. Chevalley
  1. The Betti numbers of the exceptional simple Lie groups, *Proc.*
     Intern. Congress of Math. (Cambridge, Mass., 1950), vol. 2, Amer.
     Math. Soc., Providence RI, 1952, pp. 21–24.
  2. Invariants of finite groups generated by reflections, *Amer. J. Math.*
     77 (1955), 778–782.
  3. Sur certains groupes simples, *Tôhoku Math. J.* **7** (1955), 14–66.

A.M. Cohen
  1. Finite complex reflection groups, *Ann. Sci. École Norm. Sup.* **9**
     (1976), 379–436.
  2. Finite quaternionic reflection groups, *J. Algebra* **64** (1980), 293–
     324.
  3. Coxeter groups and three related topics, lecture notes.

A.J. Coleman
  1. The Betti numbers of the simple Lie groups, *Canad. J. Math.* **10**
     (1958), 349–356.
  2. Killing and the Coxeter transformation of Kac–Moody Lie alge-
     bras, *Invent. Math.* **95** (1989), 447–477.

J.H. Conway, T.R. Curtis, S.P. Norton, R.A. Parker, R.A. Wilson
  1. *Atlas of Finite Groups*, Clarendon Press, Oxford, 1985.

M. Couillens
  1. Algèbres de Hecke, *Séminaire sur les groupes finis II*, Publ. Math.
     de l'Université Paris VII, 1983, pp. 77–94.

H.S.M. Coxeter
  1. *Regular Polytopes*, 3rd edn., Dover, New York, 1973.
  2. Discrete groups generated by reflections, *Ann. of Math.* **35** (1934),
     588–621.
  3. The complete enumeration of finite groups of the form R_i^2
     $= (R_i R_j)^{k_{ij}} = 1$, *J. London Math. Soc.* **10** (1935), 21–25.
  4. The product of the generators of a finite group generated by re-
     flections, *Duke Math. J.* **18** (1951), 765–782.
  5. Finite groups generated by unitary reflections, *Abh. Math. Sem.*
     Univ. Hamburg **31** (1967), 125–135.
  6. Regular and semi-regular polytopes III, *Math. Z.* **200** (1988), 3–45.

H.S.M. Coxeter, W.O.J. Moser
  1. *Generators and relations for discrete groups*, 3rd revised edn.,
     Springer, New York, 1980.

C.W. Curtis

1. Representations of finite groups of Lie type, *Bull. Amer. Math. Soc. (N.S.)* **1** (1979), 721–757.

2. The Hecke algebra of a finite Coxeter group, *The Arcata Conference on Representations of Finite Groups*, Proc. Symp. Pure Math. 47, part 1, Amer. Math. Soc., Providence RI, 1987, pp. 51–60.

3. Representations of Hecke algebras, *Orbites unipotentes et représentations, I. Groupes finis et algèbres de Hecke, Astérisque*, **168** (1988), pp. 13–60.

C.W. Curtis, G.I. Lehrer

1. Generic chain complexes and finite Coxeter groups, *J. Reine Angew. Math.* **363** (1985), 146–173.

C.W. Curtis, I. Reiner

1. *Representation Theory of Finite Groups and Associative Algebras*, Wiley Interscience, New York, 1962.

2. *Methods of Representation Theory* I, Wiley Interscience, New York, 1981.

3. *Methods of Representation Theory* II, Wiley Interscience, New York, 1987.

M.W. Davis

1. Groups generated by reflections and aspherical manifolds not covered by Euclidean space, *Ann. of Math.* **117** (1983), 293–324.

2. Coxeter groups and aspherical manifolds, *Algebraic topology (Aarhus, 1982)*, Lect. Notes in Math. 1051, Springer, Berlin, 1984, pp. 197–221.

3. The homology of a space on which a reflection group acts, *Duke Math. J.* **55** (1987), 97–104.

4. Some aspherical manifolds, *Duke Math. J.* **55** (1987), 105–139.

M.W. Davis, M.D. Shapiro

1. Coxeter groups are almost convex, *Geom. Dedicata* **39** (1991), 55–57.

M. Demazure

1. Désingularisation des variétés de Schubert généralisées, *Ann. Sci. École Norm. Sup.* **7** (1974), 53–88.

V.V. Deodhar

1. Some characterizations of Bruhat ordering on a Coxeter group and determination of the relative Möbius function, *Invent. Math.* **39** (1977), 187–198.

2. On Bruhat ordering and weight-lattice ordering for a Weyl group, *Indag. Math.* **40** (1978), 423–435.

190   References

3. On the Kazhdan–Lusztig conjectures, *Indag. Math.* **44** (1982), 1–17.

4. On the root system of a Coxeter group, *Comm. Algebra* **10** (1982), 611–630.

5. On some geometric aspects of Bruhat orderings. I. A finer decomposition of Bruhat cells, *Invent. Math.* **79** (1985), 499–511.

6. Local Poincaré duality and non-singularity of Schubert varieties, *Comm. Algebra* **13** (1985), 1379–1388.

7. Some characterizations of Coxeter groups, *Enseign. Math.* **32** (1986), 111–120.

8. On some geometric aspects of Bruhat orderings. II. The parabolic analogue of Kazhdan–Lusztig polynomials, *J. Algebra* **111** (1987), 483–506.

9. A splitting criterion for the Bruhat orderings on Coxeter groups, *Comm. Algebra* **15** (1987), 1889–1894.

10. A note on subgroups generated by reflections in Coxeter groups, *Arch. Math.* **53** (1989), 543–546.

11. A combinatorial setting for questions in Kazhdan–Lusztig Theory, *Geom. Dedicata* **36** (1990), 95–119.

12. Duality in parabolic set up for questions in Kazhdan–Lusztig theory, *J. Algebra* **142** (1991), 201–209.

J.M. Douglass

1. An inversion formula for relative Kazhdan–Lusztig polynomials, *Comm. Algebra* **18** (1990), 371–387.

A. Dress

1. On finite groups generated by pseudoreflections, *J. Algebra* **11** (1969), 1–5.

J. Du

1. The decomposition into cells of the affine Weyl group of type $\tilde{B}_3$, *Comm. Algebra* **16** (1988), 1383–1409.

2. Two-sided cells of the affine Weyl group of type $\tilde{C}_3$, *J. London Math. Soc.* **38** (1988), 87–98.

3. Cells in the affine Weyl group of type $\widetilde{D}_4$, *J. Algebra* **128** (1990), 384–404.

4. Sign types and Kazhdan–Lusztig cells, *Chinese Ann. Math. Ser. B* **12** (1991), 33–39.

P. DuVal

1. *Homographies, Quaternions and Rotations*, Clarendon Press, Oxford, 1964.

M. Dyer

1. Hecke algebras and reflections in Coxeter groups, PhD thesis, University of Sydney, 1987.

2. On some generalisations of the Kazhdan–Lusztig polynomials for 'universal' Coxeter groups, *J. Algebra* **116** (1988), 353–371.

3. Reflection subgroups of Coxeter systems, *J. Algebra* **135** (1990), 57–73.

4. On the "Bruhat graph" of a Coxeter system, *Compositio Math.* **78** (1991), 185–191.

5. Hecke algebras and shellings of Bruhat intervals I, to appear.

6. Hecke algebras and shellings of Bruhat intervals II: twisted Bruhat orders, to appear.

7. Quotients of twisted Bruhat orders, to appear.

M.J. Dyer, G.I. Lehrer

1. On positivity in Hecke algebras, *Geom. Dedicata* **35** (1990), 115–125.

L. Flatto

1. Basic sets of invariants for finite reflection groups, *Bull. Amer. Math. Soc.* **74** (1968), 730–734.

2. Invariants of finite reflection groups and mean value problems II, *Amer. J. Math.* **92** (1970), 552–561.

3. Invariants of finite reflection groups, *Enseign. Math.* **24** (1978), 237–292.

L. Flatto, M.M. Wiener

1. Invariants of finite reflection groups and mean value problems, *Amer. J. Math.* **91** (1969), 591–598.

J.S. Frame

1. The classes and representations of the groups of 27 lines and 28 bitangents, *Ann. Mat. Pura Appl.* (4) **32** (1951), 83–119.

A.M. Garsia, T.J. McLarnan

1. Relations between Young's natural and the Kazhdan–Lusztig representations of S_n, *Adv. in Math.* **69** (1988), 32–92.

S.I. Gelfand, R. MacPherson

1. Verma modules and Schubert cells: a dictionary, *Sém. d'algèbre P. Malliavin et M.-P. Malliavin*, Lect. Notes in Math. 925, Springer, Berlin, 1982, pp. 1–50.

D.M. Goldschmidt

1. Abstract reflections and Coxeter groups, *Proc. Amer. Math. Soc.* **67** (1977), 209–214.

M. Goresky
1. Kazhdan–Lusztig polynomials for classical groups, Northeastern University Mathematics Dept. [no date].

R.L. Griess, Jr.
1. Quotients of infinite reflection groups, *Math. Ann.* **263** (1983), 267–278.

L.C. Grove
1. The characters of the hecatonicosahedroidal group, *J. Reine Angew. Math.* **265** (1974), 160–169.

L.C. Grove, C.T. Benson
1. *Finite Reflection Groups*, 2nd edn., Springer, New York, 1985.

E. Gutkin
1. Geometry and combinatorics of groups generated by reflections, *Enseign. Math.* **32** (1986), 95–110.

2. Schubert calculus on flag varieties of Kac–Moody Lie groups, *Algebras Groups Geom.* **3** (1986), 27–59.

3. Operator calculi associated with reflection groups, *Duke Math. J.* **55** (1987), 1–18.

A. Gyoja
1. A generalized Poincaré series associated to a Hecke algebra of a finite or p-adic Chevalley group, *Japan J. Math. (N.S.)* **9** (1983), 87–111.

2. On the existence of a W-graph for an irreducible representation of a Coxeter group, *J. Algebra* **86** (1984), 422–438.

A. Gyoja, K. Uno
1. On the semisimplicity of Hecke algebras, *J. Math. Soc. Japan* **41** (1989), 75–79.

Z. Haddad
1. Infinite dimensional flag varieties, PhD thesis, M. I. T., 1984.

2. A Coxeter group approach to Schubert varieties, *Infinite-dimensional groups with applications*, Springer, New York, 1985, pp. 157–165.

P. de la Harpe
1. Groupes de Coxeter infinis non affines, *Exposition. Math.* **5** (1987), 91–96.

2. An invitation to Coxeter groups, *Group Theory from a Geometrical Viewpoint* (Trieste, 1990), World Scientific, Singapore, 1991.

M. Hazewinkel, W. Hesselink, D. Siersma, F.D. Veldkamp
1. The ubiquity of Coxeter–Dynkin diagrams (an introduction to the A-D-E problem), *Nieuw Arch. Wisk.* **25** (1977), 257–307.

A. Heck

1. A criterion for a triple (X, I, μ) to be a W-graph of a Coxeter group, *Comm. Algebra* **16** (1988), 2083–2102.

H.L. Hiller

1. Schubert calculus of a Coxeter group, *Enseign. Math.* **27** (1981), 57–84.
2. Combinatorics and intersections of Schubert varieties, *Comment. Math. Helv.* **57** (1982), 41–59.
3. *Geometry of Coxeter Groups*, Research Notes in Mathematics, No. 54, Pitman, Boston, 1982.

M.E. Hoffman, W.D. Withers

1. Generalized Chebyshev polynomials associated with affine Weyl groups, *Trans. Amer. Math. Soc.* **308** (1988), 91–104.

A. van den Hombergh

1. About the automorphisms of the Bruhat-ordering in a Coxeter group, *Indag. Math.* **36** (1974), 125–131.

R.B. Howlett

1. Normalizers of parabolic subgroups of reflection groups, *J. London Math. Soc.* **21** (1980), 62–80.
2. Coxeter groups and M-matrices, *Bull. London Math. Soc.* **14** (1982), 137–141.
3. On the Schur multipliers of Coxeter groups, *J. London Math. Soc.* **38** (1988), 263–276.

R.B. Howlett, G.I. Lehrer

1. Duality in the normalizer of a parabolic subgroup of a finite Coxeter group, *Bull. London Math. Soc.* **14** (1982), 133–136.

J.E. Humphreys

1. *Introduction to Lie Algebras and Representation Theory*, Springer, New York, 1972.
2. *Linear Algebraic Groups*, Springer, New York, 1975.

B. Huppert

1. Zur Konstruktion der reellen Spiegelungsgruppe H_4, *Acta Math. Acad. Sci. Hungar.* **26** (1975), 331–336.

V.F. Ignatenko

1. Some questions in the geometric theory of invariants generated by orthogonal and oblique reflections, *J. Soviet Math.* **33** (1986), 933–953.

S. Ihara, T. Yokonuma

1. On the second cohomology groups (Schur-multipliers) of finite reflection groups, *J. Fac. Sci. Univ. Tokyo Sect. I* **11** (1965), 155–171.

H.-C. Im Hof

1. A class of hyperbolic Coxeter groups, *Exposition. Math.* **3** (1985), 179–186.

N. Iwahori

1. On the structure of a Hecke ring of a Chevalley group over a finite field, *J. Fac. Sci. Univ. Tokyo Sect. I* **10** (1964), 215–236.

N. Iwahori, H. Matsumoto

1. On some Bruhat decomposition and the structure of the Hecke rings of p-adic Chevalley groups, *Inst. Hautes Études Sci. Publ. Math.* **25** (1965), 5–48.

J.C. Jantzen

1. *Moduln mit einem höchsten Gewicht*, Lect. Notes in Math. 750, Springer, Berlin, 1979.

2. *Einhüllende Algebren halbeinfacher Lie-Algebren*, Springer, Berlin, 1983.

3. *Representations of Algebraic Groups*, Academic Press, Orlando, 1987.

V.G. Kac

1. *Infinite Dimensional Lie Algebras*, Birkhäuser, Boston, 1983; 2nd edn., Cambridge University Press, 1985.

V.G. Kac, D.H. Peterson

1. Generalized invariants of groups generated by reflections, *Geometry Today*, Birkhäuser, Boston, 1985, pp. 231–249.

V.G. Kac, K. Watanabe

1. Finite linear groups whose ring of invariants is a complete intersection, *Bull. Amer. Math. Soc. (N.S.)* **6** (1982), 221–223.

M. Kaneda

1. On the inverse Kazhdan–Lusztig polynomials for affine Weyl groups, *J. Reine Angew. Math.* **381** (1987), 116–135.

G. Karpilovsky

1. *The Schur Multiplier*, Clarendon Press, Oxford, 1987.

S.I. Kato

1. A realization of irreducible representations of affine Weyl groups, *Indag. Math.* **45** (1983), 193–201.

2. On the Kazhdan–Lusztig polynomials for affine Weyl groups, *Adv. in Math.* **55** (1985), 103–130.

D. Kazhdan, G. Lusztig

1. Representations of Coxeter groups and Hecke algebras, *Invent. Math.* **53** (1979), 165–184.

2. Schubert varieties and Poincaré duality, *Geometry of the Laplace operator*, Proc. Sympos. Pure Math. 34, Amer. Math. Soc., Providence, RI, 1980, pp. 185–203.

S.V. Kerov

1. *W*-graphs of representations of symmetric groups, *J. Soviet Math.* **28** (1985), 596–605.

D. Kletzing

1. *Structure and representations of Q-groups*, Lect. Notes in Math. 1084, Springer, Berlin, 1984.

T. Kondo

1. The characters of the Weyl group of type F_4, *J. Fac. Sci. Univ. Tokyo* **11** (1965), 145–153.

B. Kostant

1. The principal three-dimensional subgroup and the Betti numbers of a complex simple Lie group, *Amer. J. Math.* **81** (1959), 973–1032.

J.L. Koszul

1. *Lectures on hyperbolic Coxeter groups*, University of Notre Dame Math. Dept., 1967.

F. Lannér

1. On complexes with transitive groups of automorphisms, *Comm. Sém. Math. Univ. Lund* **11** (1950), 71 pp.

A. Lascoux, M.-P. Schützenberger

1. Polynômes de Kazhdan & Lusztig pour les grassmanniennes, *Young tableaux and Schur functions in algebra and geometry, Astérisque*, **87–88** (1981), pp. 249–266.

G. Lawton

1. Two-sided cells in the affine Weyl group of type $\tilde{A}_{n-1}$, *J. Algebra* **120** (1989), 74–89.

G.I. Lehrer

1. A survey of Hecke algebras and the Artin braid groups, *Braids (Santa Cruz, CA, 1986), Contemp. Math.* **78**, Amer. Math. Soc., Providence, RI, 1988, pp. 365–385.

2. On the Poincaré series associated with Coxeter group actions on the complements of hyperplanes, *J. London Math. Soc.* **36** (1987), 275–294.

G.I. Lehrer, L. Solomon

1. On the action of the symmetric group on the cohomology of the complement of its reflecting hyperplanes, *J. Algebra* **104** (1986), 410–424.

G. Lusztig

1. Some problems in the representation theory of finite Chevalley groups, *The Santa Cruz Conference on Finite Groups*, Proc. Sympos. Pure Math. 37, Amer. Math. Soc., Providence, RI, 1980, pp. 313–317.

2. Hecke algebras and Jantzen's generic decomposition patterns, *Adv. in Math.* **37** (1980), 121–164.

3. Some examples of square integrable representations of semisimple *p*-adic groups, *Trans. Amer. Math. Soc.* **277** (1983), 623–653.

4. Left cells in Weyl groups, *Lie Group Representations I*, Lect. Notes in Math. 1024, Springer, Berlin, 1984, pp. 99–111.

5. *Characters of reductive groups over a finite field*, Ann. of Math. Studies 107, Princeton Univ. Press, 1984.

6. Cells in affine Weyl groups, *Algebraic Groups and Related Topics*, Adv. Studies in Pure. Math. 6, North-Holland, Amsterdam, 1985, pp. 225–287.

7. The two-sided cells of the affine Weyl group of type A_n, *Infinite-dimensional groups with applications*, Springer, New York, 1985, pp. 275–283.

8. Sur les cellules gauches des groupes de Weyl, *C.R. Acad. Sci. Paris. Sér. I Math.* **302** (1986), 5–8.

9. Cells in affine Weyl groups II, *J. Algebra* **109** (1987), 536–548.

10. Cells in affine Weyl groups III, *J. Fac. Sci. Univ. Tokyo Sect. IA Math.* **34** (1987), 223–243.

11. Cells in affine Weyl groups IV, *J. Fac. Sci. Univ. Tokyo Sect. IA Math.* **36** (1989), 297–328.

G. Lusztig, Xi Nanhua

1. Canonical left cells in affine Weyl groups, *Adv. in Math.* **72** (1988), 284–288.

I.G. Macdonald

1. *Spherical functions on a group of p-adic type*, Publ. Ramanujan Inst. No. 2, Madras, 1971.

2. The Poincaré series of a Coxeter group, *Math. Ann.* **199** (1972), 161–174.

3. Affine root systems and Dedekind's η-function, *Invent. Math.* **15** (1972), 91–143.

4. On the degrees of the irreducible representations of finite Coxeter groups, *J. London Math. Soc.* **6** (1973), 298–300.

5. Affine Lie algebras and modular forms, *Sém. Bourbaki (1980/81)*, Exp. 577, Lect. Notes in Math. 901, Springer, Berlin, 1982, pp. 258–276.

D. Martinais

1. Classification des groupes cristallographiques associés aux groupes de Coxeter irréductibles, *C.R. Acad. Sci. Paris Sér. I Math.* **302** (1986), 335–338.

H. Matsumoto

1. Générateurs et relations des groupes de Weyl généralisés, *C.R. Acad. Sci. Paris* **258** (1964), 3419–3422.

G. Maxwell

1. The crystallography of Coxeter groups, *J. Algebra* **35** (1975), 159–177.

2. On the crystallography of infinite Coxeter groups, *Math. Proc. Cambridge Philos. Soc.* **82** (1977), 13–24.

3. The Schur multipliers of rotation subgroups of Coxeter groups, *J. Algebra* **53** (1978), 440–451.

4. Sphere packings and hyperbolic reflection groups, *J. Algebra* **79** (1982), 78–97.

5. Wythoff's construction for Coxeter groups, *J. Algebra* **123** (1989), 351–377.

M.L. Mehta

1. Basic sets of invariant polynomials for finite reflection groups, *Comm. Algebra* **16** (1988), 1083–1098

V.F. Molchanov

1. Poincaré polynomials of representations of finite groups generated by reflections, *Math. Notes* **31** (1982), 423–427.

B. Monson

1. The Schläflian of a crystallographic Coxeter group, *C.R. Math. Rep. Acad. Sci. Canada* **4** (1982), 145–147.

2. Simplicial quadratic forms, *Canad. J. Math.* **35** (1983), 101–116.

G. Moussong

1. Hyperbolic Coxeter groups, Ph.D. thesis, Ohio State University, 1988.

H. Nakajima

1. Invariants of finite groups generated by pseudoreflections in positive characteristics, *Tsukuba J. Math.* **3** (1979), 109–122.

Nguyen Viet Dung

1. The fundamental groups of the spaces of regular orbits of the affine Weyl groups, *Topology* **22** (1983), 425–435.

P. Orlik, L. Solomon

1. Unitary reflection groups and cohomology, *Invent. Math* **59** (1980), 77–94.

2. Combinatorics and topology of complements of hyperplanes, *Invent. Math.* **56** (1980), 167–189.

3. Complexes for reflection groups, *Algebraic Geometry*, Lect. Notes in Math. 862, Springer, Berlin, 1981, pp. 193–207.

4. Coxeter arrangements, *Singularities*, Part 2, Proc. Sympos. Pure Math. 40, Amer. Math. Soc., Providence, RI, 1983, pp. 269–291.

5. Arrangements defined by unitary reflection groups, *Math. Ann.* **261** (1982), 339–357.

6. The Hessian map in the invariant theory of reflection groups, *Nagoya Math. J.* **109** (1988), 1–21.

7. Discriminants in the invariant theory of reflection groups, *Nagoya Math. J.* **109** (1988), 23–45.

P. Orlik, L. Solomon, H. Terao

1. On Coxeter arrangements and the Coxeter number, *Complex analytic singularities*, Adv. Stud. Pure Math. 8, North-Holland, Amsterdam, 1987, pp. 461–477.

L. Paris

1. Growth series of Coxeter groups, *Group Theory from a Geometrical Viewpoint* (Trieste, 1990), World Scientific, Singapore, 1991.

H. Pfeuffer

1. Über die reelle Spiegelungsgruppe H_4 und die Klassenzahl der sechsdimensionalen Einheitsform, *Arch. Math.* **31** (1978/79), 126–132.

S.J. Pride, R. Stöhr

1. The co(homology) of aspherical Coxeter groups, *J. London Math. Soc.* **42** (1990), 49–63.

R.A. Proctor

1. Classical Bruhat orders and lexicographic shellability, *J. Algebra* **77** (1982), 104–126.

R.W. Richardson

1. Conjugacy classes of involutions in Coxeter groups, *Bull. Austral. Math. Soc.* **26** (1982), 1–15.

M. Ronan

1. *Lectures on Buildings*, Academic Press, San Diego, 1989.

W. Rudin

1. Proper holomorphic maps and finite reflection groups, *Indiana U. Math. J.* **31** (1982), 701–720.

K. Saito, T. Yano, J. Sekiguchi

1. On a certain generator system of the ring of invariants of a finite reflection group, *Comm. Algebra* **8** (1980), 373–408.

J. Sekiguchi, T. Yano

1. The algebra of invariants of the Weyl group $W(F_4)$, *Sci. Rep. Saitama Univ. Ser. A* **9** (1979), no. 2, 21–32.

2. A note on the Coxeter group of type H_3, *Sci. Rep. Saitama Univ. Ser. A* **9** (1979), no. 2, 33–44.

O.P. Shcherbak

1. Wavefronts and reflection groups, *Russian Math. Surveys* **43** (1988), 149–194.

G.C. Shephard

1. Unitary groups generated by reflections, *Canad. J. Math.* **5** (1953), 364–383.

2. Some problems on finite reflection groups, *Enseign. Math.* **2** (1956), 42–48.

G.C. Shephard, J.A. Todd

1. Finite unitary reflection groups, *Canad. J. Math.* **6** (1954), 274–304.

Shi Jian-yi

1. *The Kazhdan–Lusztig cells in certain affine Weyl groups*, Lect. Notes in Math. 1179, Springer, Berlin, 1986.

2. A two-sided cell in an affine Weyl group I, *J. London Math. Soc.* **36** (1987), 407–420.

3. A two-sided cell in an affine Weyl group II, *J. London Math. Soc.* **37** (1988), 253–264.

4. A result on the Bruhat order of a Coxeter group, *J. Algebra* **128** (1990), 510–516.

L. Smith

1. On the invariant theory of finite pseudo reflection groups, *Arch. Math.* **44** (1985), 225–228.

L. Solomon

1. Invariants of finite reflection groups, *Nagoya Math. J* **22** (1963), 57–64.

2. Invariants of Euclidean reflection groups, *Trans. Amer. Math. Soc.* **113** (1964), 274–286.

3. The orders of the finite Chevalley groups, *J. Algebra* **3** (1966), 376–393.

4. A decomposition of the group algebra of a finite Coxeter group, *J. Algebra* **9** (1968), 220–239.

5. A Mackey formula in the group ring of a Coxeter group, *J. Algebra* **41** (1976), 255–264.

200   References

T.A. Springer

   1. Some arithmetical results on semi-simple Lie algebras, *Inst. Hautes Études Sci. Publ. Math.* **30** (1966), 115–141.

   2. Regular elements of reflection groups, *Invent. Math.* **25** (1974), 159–198.

   3. *Invariant Theory*, Lect. Notes in Math. 585, Springer, Berlin, 1977.

   4. A construction of representations of Weyl groups, *Invent. Math.* **44** (1978), 279–293.

   5. Some remarks on involutions in Coxeter groups, *Comm. Algebra* **10** (1982), 631–636.

   6. Quelques applications de la cohomologie d'intersection, *Sém. Bourbaki (1981/82)*, Exp. 589, *Astérisque* **92–93** (1982).

R.P. Stanley

   1. Relative invariants of finite groups generated by pseudoreflections, *J. Algebra* **49** (1977), 134–148.

   2. Invariants of finite groups and their applications to combinatorics, *Bull. Amer. Math. Soc. (N.S.)* **1** (1979), 475–511.

   3. Weyl groups, the hard Lefschetz theorem, and the Sperner property, *SIAM J. Alg. Disc. Meth.* **1** (1980), 168–184

   4. On the number of reduced decompositions of elements of Coxeter groups, *European J. Combin.* **5** (1984), 359–372.

R. Steinberg

   1. Finite reflection groups, *Trans. Amer. Math. Soc.* **91** (1959), 493–504.

   2. Invariants of finite reflection groups, *Canad. J. Math.* **12** (1960), 616–618.

   3. Differential equations invariant under finite reflection groups, *Trans. Amer. Math. Soc.* **112** (1964), 392–400.

   4. *Lectures on Chevalley Groups*, Yale University Math. Dept., 1968.

   5. *Endomorphisms of linear algebraic groups*, Mem. Amer. Math. Soc. No. 80 (1968).

   6. Finite subgroups of SU_2, Dynkin diagrams and affine Coxeter elements, *Pacific J. Math.* **118** (1985), 587–598.

E. Stiefel

   1. Über eine Beziehung zwischen geschlossenen Lie'schen Gruppen und diskontinuierlichen Bewegungsgruppen euklidischer Räume und ihre Anwendung auf die Aufzählung der einfachen Lie'schen Gruppen, *Comment. Math. Helv.* **14** (1941/42), 350–380.

E. Straume

   1. The topological version of groups generated by reflections, *Math. Z.* **176** (1981), 429–445.

K. Takahashi

1. The left cells and their W-graphs of Weyl group of type F_4, *Tokyo J. Math.* **13** (1990), 327–340.

H. Terao

1. Generalized exponents of a free arrangement of hyperplanes and Shephard–Todd–Brieskorn formula, *Invent. Math.* **63** (1981), 159–179.

J. Tits

1. Groupes et géométries de Coxeter, IHES, 1961.

2. Racines et groupes engendrés par reflexions, *Algébres de Lie semisimples et systèmes de racines*, Université Libre de Bruxelles, 1963–1964.

3. Géométries polyédriques finies, *Rend. Mat. e Appl.* **23** (1964), 156–165.

4. Structures et groupes de Weyl, *Sém. Bourbaki (1964/65)*, Exp. 288, Secrétariat Mathématique, Paris, 1966.

5. Le problème des mots dans les groupes de Coxeter, *Symposia Mathematica (INDAM, Rome, 1967/68)*, Academic Press, London, 1969, vol. 1, pp. 175–185.

6. *Buildings of spherical type and finite BN-pairs*, Lect. Notes in Math. 386, Springer, Berlin, 1974.

7. Two properties of Coxeter complexes, *J. Algebra* **41** (1976), 265–268.

8. Endliche Spiegelungsgruppen, die als Weylgruppen auftreten, *Invent. Math.* **43** (1977), 283–295.

9. Sur le groupe des automorphismes de certains groupes de Coxeter, *J. Algebra* **113** (1988), 346–357.

T. Uzawa

1. Finite Coxeter groups and their subgroup lattices, *J. Algebra* **101** (1986), 82–94.

D.-N. Verma

1. Möbius inversion for the Bruhat ordering on a Weyl group, *Ann. Sci. École Norm. Sup.* **4** (1971), 393–398.

2. A strengthening of the exchange property of Coxeter groups, preprint, 1972.

3. The rôle of affine Weyl groups in the representation theory of algebraic Chevalley groups and their Lie algebras, *Lie groups and their representations*, Halsted, New York, 1975, pp. 653–705.

E.B. Vinberg

1. Discrete groups generated by reflections in Lobacevskii spaces, *Math. USSR-Sb.* **1** (1967), 429–444.

2. Geometric representations of Coxeter groups [Russian], *Uspekhi Mat. Nauk* **25** (1970), no. 2, 267–268.

3. Discrete linear groups generated by reflections, *Math. USSR–Izv.* **5** (1971), 1083–1119.

4. Discrete reflection groups in Lobachevsky spaces, *Proc. Intern. Congress Math. (Warsaw, 1983)*, PWN, Warsaw, 1984, pp. 593–601.

5. Hyperbolic reflection groups, *Russian Math. Surveys* **40** (1985), 31–75.

D.A. Vogan, Jr
  1. A generalized τ-invariant for the primitive spectrum of a semisimple Lie algebra, *Math. Ann.* **242** (1979), 209–224.

P. Wagreich
  1. The growth function of a discrete group, *Group actions and vector fields*, Lect. Notes in Math. 956, Springer, Berlin, 1982, pp. 125–144.

W.C. Waterhouse
  1. Automorphisms of the Bruhat order on Coxeter groups, *Bull. London Math. Soc.* **21** (1989), 243–248.

E. Witt
  1. Spiegelungsgruppen und Aufzählung halbeinfacher Liescher Ringe, *Abh. Math. Sem. Univ. Hamburg* **14** (1941), 289–322.

Xi Nanhua
  1. An approach to the connectedness of the left cells in affine Weyl groups, *Bull. London Math. Soc.* **21** (1989), 557–561.

T. Yokonuma
  1. On the second cohomology groups (Schur multipliers) of infinite discrete reflection groups, *J. Fac. Sci. Univ. Tokyo Sect. I* **11** (1965), 173–186.

Index